SpringerBriefs in Earth Sciences

Series Editor

Pedro Maciel de Paula Garcia, Universidade Federal de Mato Grosso
Cuiabá, Brazil

SpringerBriefs in Earth Sciences present concise summaries of cutting-edge research and practical applications in all research areas across earth sciences. It publishes peer-reviewed monographs under the editorial supervision of an international advisory board with the aim to publish 8 to 12 weeks after acceptance. Featuring compact volumes of 50 to 125 pages (approx. 20,000–70,000 words), the series covers a range of content from professional to academic such as:

- timely reports of state-of-the art analytical techniques
- bridges between new research results
- snapshots of hot and/or emerging topics
- literature reviews
- in-depth case studies

Briefs will be published as part of Springer's eBook collection, with millions of users worldwide. In addition, Briefs will be available for individual print and electronic purchase. Briefs are characterized by fast, global electronic dissemination, standard publishing contracts, easy-to-use manuscript preparation and formatting guidelines, and expedited production schedules.

Both solicited and unsolicited manuscripts are considered for publication in this series.

Hannu Ahokas • Elizabeth L. Chamberlain
Mark D. Bateman

Paleoshorelines and Their Relict Littoral Plants

Tracing the Past Baltic Sea Shorelines and Discharge of the Kymi River

Springer

Hannu Ahokas
Helsinki, Finland

Mark D. Bateman
School of Geography and Planning
University of Sheffield
Sheffield, UK

Elizabeth L. Chamberlain
Soil Geography & Landscape Group
Wageningen University & Research
Wageningen, The Netherlands

ISSN 2191-5369 ISSN 2191-5377 (electronic)
SpringerBriefs in Earth Sciences
ISBN 978-3-032-17722-3 ISBN 978-3-032-17723-0 (eBook)
https://doi.org/10.1007/978-3-032-17723-0

This Springer imprint is published by the registered company Springer Nature Switzerland AG
The registered company address is: Gewerbestrasse 11, 6330 Cham, Switzerland

Preface

This topic evolved while I worked since my childhood in the field of the Saarela farm, Elimäki municipality, southeastern Finland. Since the age of 4 years in 1951, I had to fill a man's position at work with a stationary threshing machine. I encountered puzzling occurrences of reed (*Phragmites australis*) in the arable field. The stems of reed were well observable in the sheaves of crops. Reed is normally a seaside species. It was obvious that in prehistory, there had been a shore from which the reed was a littoral relic in the field. No such shore level was recognized in the literature, which knew the Ancylus Lake as the last one to submerge the area by about 20 m. Two reed stands occupied 0.86 and 1.51 ha in the field. For the dating, I sought ancient, raised shorelines around the Saarela farm, the nearest clear paleo-shores being 1.88 km ESE on the same level, on the sandy flanks of a north-south directed esker. The dated and now published ages of the littoral stage are novel. The paleo-shore sands here have usually evolved from prior glaciofluvial deposits, which accumulated on stream bottoms during the melting of the Ice Age. The presented ages of 63 ± 3 and 56 ± 2.7 ka reveal that a glaciofluvial stream occurred during the Middle Weichselian melting. After the melting, mammoths could survive here.

The land of this region rises a few millimetres annually. Additionally, sea level changes of the Baltic Sea have left a number of shorelines in the region. Towards the end of the Ice Age, during chilled climate epochs, major ridges were formed at the ice margins during the stop of the retreating ice. The outermost of the ridges is officially called the First Salpausselkä in Finland. The descriptive appellative *salpa* denotes 'dam', 'blockage', and 'bar'. Only three rivers had burst out through the First Salpausselkä. The outbursts were in part a consequence of the uneven land upheaval, which in effect tilted the land of Finland-to-be towards the southeast: this also raises the question of whether the most southeastern of the three rivers, the Vuoksi, was the first to outburst the First Salpausselkä ridge. The inland occurrences of four relict, originally littoral plant species, chive (*Allium schoenoprasum*), sheep's-bit (*Jasione montana*), trefoil (*Lotus corniculatus*), and spiked speedwell (*Veronica spicata*), were related to the past littoral levels. The relict species originating from warmer climate phases now occur sporadically on hill verges, typically on the southern exposition. The relict plants seem to have maintained the altitude of the ancient littoral zone, surviving in sites that meet some ecological key requirements. The inland chive has survived on gently sloping rocks receiving seeping moisture, which maintains cyanobacterial biofilm on the rock. The cyanobacteria fix nitrogen and the nitrogen leaks after intermittent droughts to the nearby chives, which are highly dependent on the nitrogen, and with their bulbs, survive the droughts.

The river Kymi was a boundary of the historical Elimäki, where the folk tradition knew that 'Elimäki is part of the great bottom of the Kymi'. The outburst of the Kymi through the First Salpausselkä ridge caused vast floods in Elimäki and formed sand plains now dated with the luminescence method to 7.61 ± 0.46 ka ago, which is also the first direct dating for the Kymi discharge. Radiocarbon dates suggest that flooding was frequent and reached many areas. Supercooling of water is known to have caused blockage of the Kymi, leading to fast-rising floods. Floods in the Kymi river are known from oral traditions, documentary records, dating, and toponyms. Floods, rainy, and cool climate periods caused the development of swamps. Tens of toponyms carry appellatives referring to swamps. Many terrain points staying above the floods carry an appellative denoting an island.

Prehistoric paintings on upright rocks were typically made standing on the ice of the adjoining lake. An age estimate is presented for a prehistoric painting in Elimäki.

Helsinki, Finland
January, 2025

Hannu Ahokas

Acknowledgements

Alice Versendaal (NCL, Wageningen, The Netherlands) and Rob Ashurst (Sheffield University, UK) are thanked for technical support in the study. The staff of Beta Analytic Inc. (Miami, FL, USA) is thanked for the radiocarbon dating, the results of which are published now for the first time. The investment grant (#834.03.003) supplied by the Netherlands Organization for Scientific Research (NOW-ALW) to the Dutch collaborators is acknowledged.

Competing Interests The authors have no competing interests to declare that are relevant to the content of this manuscript.

About the Book

Ancient Fennoscandian coastlines have been exposed through the Holocene, primarily due to glacial isostatic adjustment, resulting in relict inland shorelines that are detectable at elevations significantly higher than present-day sea levels. These paleoshorelines provide insight into past sea levels and regional paleoenvironments. In this book, previously unmapped paleoshoreline elevations between Ancylus Lake and the Litorina Sea of the Baltic Basin are examined along the southeastern coast of Finland. Using a novel combination of geochronological and paleoenvironmental indicators, including relict littoral plant species, shipwrecks, prehistoric rock art, luminescence, and radiocarbon dating, we offer a reconstruction of the paleoenvironment in the study area. Prehistoric rock art and shipwrecks can serve as indicators of sea levels. The glaciofluvial sands of the Välikangas esker in S Kouvola were luminescence dated to 63 ± 3 and 56 ± 2.7 ka, which represent the Middle Weichselian melting. Stormy weather between 11.5 ± 0.4 and 9.5 ± 0.4 ka ago caused sand redeposition on the sides of glaciofluvial eskers. These redeposited sands preserved evidence of paleoshorelines, which appear as raised beaches at elevations of 35–43 m above sea level (asl). These features were dated to between 7.4 and 3.8 cal ka BP. An outburst of floodwater southwards through an esker formed the Koria sand plains, dated by luminescence to 7.61 ± 0.46 ka ago, which also provides an age estimate for the outburst of the Kymi River through the First Salpausselkä ridge. Major streaming in the Koria sand plains between 3.65 and 2.0 cal ka BP is also suggested by radiocarbon dating. More recent floods caused by ice blockages of rapids in the Kymi River are documented in oral traditions, documentary records, dating, and toponyms. These floods are thought to have followed cold or rainy epochs between 2.3 and 1.9 cal ka BP, at *fimbulvinteren c.* 1.41 cal ka BP, and during the Little Ice Age at *c.* 350 BP.

Contents

About the Authors

Hannu Ahokas is a plant geneticist and retired research specialist from Agrifood Research Finland. He obtained his Ph.D. degree from the University of Helsinki in 1984. His non-professional botanical and geological excursions and sample collecting for over 30 years led to the explorations dealt with in the book via international collaboration. As an output in genetics, he detected in 1976 a cytoplasmic male sterility system in barley enabling mass hybridization, which has been commercialized to produce winter barley hybrid cultivars in Europe since 2002.

Elizabeth L. Chamberlain is an Assistant Professor at Wageningen University, the Netherlands, and the 2021 International Association of Sedimentologists awardee for outstanding work in marine sedimentology. She was educated at Tulane University and Louisiana State University, both situated in the beautiful Mississippi Delta, USA. Her ongoing work explores geomorphology and human–landscape dynamics in coasts and deltas worldwide, including those of Bangladesh, Peru, and the Netherlands.

Mark D. Bateman is a Professor in Palaeoenvironments at the University of Sheffield, where he has conducted research for the last 30 years. He has a B.Sc. in Geography from the University of London and a D.Phil. from the University of Sussex. He has over 200 publications in his research area of the use of sediment archives and luminescence dating for a better understanding of past depositional processes and environmental changes. In 2018, he was awarded the Sorby Medal by the Yorkshire Geological Society.

Abbreviations

AD	Anno Domini
AMS	Accelerator mass spectrometry
asl	Above sea level
c.	circa
cal a BP	Calibrated years before the calendar year 1950 or Before Present
cal ka BP	Calibrated 1000 years before the calendar year 1950 or Before Present
ka	1000 years
p	Probability
uncal	Uncalibrated

Chapter 1
Introduction

Abstract Elements of today's landscapes, including geomorphology, ecology, and even cultural artifacts of many regions, allow for reconstructing past environments that are no longer visible. This book explores the ancient landscape of Finland around the Baltic Sea, as revealed through floral communities, sediments, shipwrecks, and rock art. Once located where the land met the sea, these littoral archives are now isolated from the waters that yielded them. From the perspective of a Finnish plant geneticist and geologists, and supplemented with cutting edge dating tools, you will learn about the paleoenvironmental clues left by the changing Baltic coastline.

Keywords Archaeology · Finland · Floral communities · Luminescence dating · Paleoenvironment · Sea level

Elements of today's landscapes, including geomorphology, ecology, and even cultural artifacts of many regions, allow for reconstructing past environments that are no longer visible. This book explores the ancient landscape of Finland around the Baltic Sea, as revealed through floral communities, sediments, shipwrecks, and rock art. Once located where the land met the sea, these littoral archives are now isolated from the waters that yielded them. The land that encountered upheaval shows signs of a gradual advance of the shoreline away from the land. In the dynamic Baltic Sea region, an ancient land that has undergone upheaval as glaciers melted shows signs of a gradual retreat of the seashore from the land, despite rising global sea levels.

Plant communities are intimately linked to geomorphology and local water chemistry. On muddy sites, grass and forest vegetation often appeared after the sea waves retreated. On rocky shore sites, waves tended to keep the surface clear of bigger plants. On such open rocks with cracks and some gravel, the resistant shore vegetation may stay for years, centuries, or millennia after the sea has withdrawn, if other ecological requirements are met. A vast gene pool of the seashore species

H. Ahokas et al., *Paleoshorelines and Their Relict Littoral Plants*, SpringerBriefs in Earth Sciences, https://doi.org/10.1007/978-3-032-17723-0_1

can help the adaptation to conditions deviating from those of the seaside. The seashore plant species tolerate sunshine, also reflected from the water, or are heliotropic, requiring plenty of sunshine. Therefore, the relic seashore species in inland sites most likely survived on southern exposures, which receive the most sun, in the northern latitudes around the Baltic Sea. After land upheaval, shallow underwater depressions may continue as separated lakes from the sea, or as peat-forming swamps. The widespread seashore species, reed (*Phragmites australis*), survives long in such engulfed lakes, but less well in peaty swamps. As the peat layer develops, the acidification of the swamp hinders the growth of the reed. Reed typically occurs in nearly neutral or even basic habitats, with a pH range of up to 9.5, and predominantly above pH 5.6 (Björk 1967). Bottoms with clay minerals rich with basic cations make the occurrence of reed possible in swamps (Kutschera 1960).

Around the Baltic Sea, in the Baltic Basin, another type of factor positioned new shores even hundreds of kilometers from the current shores towards the inland. In the eustatic sea level changes, the water settled to a new, usually lower level. Sea level rises, transgressions to new levels could also have occurred (Jantunen and Donner 1996). The changes in the sea level were typically followed by changes in the edaphic properties, or salt content of the water. Such edaphic changes occurred unevenly in different places and depths and apparently slowly in the Baltic. Overall, the macroclimate fluctuated. In such conditions, a niche sometimes remained that could sustain a seashore plant species as an inland relic.

An example of a rarely surviving seashore species is the chive (*Allium schoenoprasum*), which is common on the open, gravelly, and rocky shores of the present-day Baltic coast and islands. The survival of the nitrophilous chive on inland rocks of past shorelines seems to be linked to the presence of nitrogen-fixing cyanobacteria on the open rock. Good growth of the cyanobacteria requires a source of seeping water to continue the wetting of the rock surface between the rains. On the other hand, chives can survive dry periods with their onions, while other vascular species are weakened and eliminated by the drought, which keeps the habitat open and sunny, favorable for chives (Ahokas 2018).

Many of the sand and gravel resources of Finland were mapped to have been formed by glaciofluvial rivers during the Ice Age, but were undated. In the research for this book, we had two samples that dated the stratification of sand in an esker to the middle, melt era of the latest Ice Age. This melting era must have permitted the growth of grass and shrubs, which made the life of the mammoths possible here. As a big animal, a mammoth needed plenty of grass-fed daily throughout the year, which required ice-free land. The mammoth bones found in Finland and dated show a melting period to have occurred during the Ice Age (Ukkonen et al. 1999; Ukkonen and Mannermaa 2017).

After the Ice Age, outbursts of new rivers were associated with land upheaval, land tilting, and eskers blocking newly formed lakes. The pioneer inhabitants could see such outbursts and floods of the waters, ultimately eroding new river channels. Intervening ice blockages enforced the succeeding floods here in the sub-arctic north. Supercooled water could form a tight ice blockage down to the bottom in a short time. Throughout this volume, we explore in detail littoral plants of Southeastern Finland and environmental conditions past and present that have supported them as recorded through various human and natural sources. We make an effort to date the outbursts and floods of a major river since the end of the Ice Age, which have affected human life since prehistory. Varied archives are used as tools to uncover the obscured ancient landscape of Southeastern Finland.

1.1 General Description of the Research Region in Southeastern Finland

The research region comprises the southeastern coast of Finland, approximately 67 km in N–S direction between the Second Salpausselkä end moraine ridge (Ss2) and the Gulf of Finland. The area extends approximately 86 km in the east-west direction, with the Kymi (Kymijoki) River at its center and a few observation sites located beyond these limits. The region has encountered all the ice-free sea stages of the Baltic Basin after the Weichselian Ice Age: Baltic Ice Lake, Yoldia Sea, and Ancylus Lake, and the lands below *c.* 30 m asl, also Litorina Sea (Miettinen 2002; Tikkanen and Oksanen 2002; Ancient shorelines 2013; Ojala et al. 2013). The region of the study lies on the Pre-Quaternary massive of rapakivi granites (Simonen and Lehijärvi 1963; Simonen and Laitala 1970); for the geology of the local rapakivi granites, see e.g. the article of Rämö et al. (2014). The depth of the Quaternary deposits in the region is mostly between 10 and 30 m, but significantly deeper in the two Salpausselkä ridges, Ss1 and Ss2 (Fig. 1.1), with bare rocks forming a minority of the surface (GTK maankamara 2024). Clay deposits reach in the northern region up to 80 m asl, and in the south, 40 m asl (Valovirta 1972; Sahala 1988), and comprise roughly 40% of the surface (Tynni et al. 1976; Rainio et al. 1989a, b). Glaciofluvial deposits formed eskers, which were modified by later ice advances and now start in the region from Ss1, continuing in an SSE direction. The prehistoric direction downstream towards the sea, currently the Gulf of Finland, has evidently been retained for over 60 ka. On their sandy flanks, these eskers often have preserved raised beach lines from the different later stages of the Baltic Sea.

The surface clays were typically deposited after the glaciofluvial eskers were formed making the clay—esker-sand contact level along each esker a characteristic of the land tilting in the land upheaval (Fig. 1.1). Besides the major Kymi river and the two rivers eastwards (details in the Sect. 4.7), which discharged the Ss1, other rivers flow in the region from Ss1 to the Gulf of Finland. Lakes are frequent north of the Ss1 and rarer south of it. The continuations of the Ss1, Ss2, and Ss3 ridges were recently depicted in the whole Fennoscandian extent by Boyes et al. (2024).

The aim of this research was to reconstruct post-glacial landscape evolution in terms of former shorelines and fluvial activity using a novel combination of geochronological and paleoenvironmental indicators, including relict littoral plant species, raised beaches, luminescence dating, and radiocarbon (^{14}C) dating. Archaeological and historical data, as well as ignored but recorded oral traditions and onomastics, are presented to substantiate the geomorphic and environmental changes.

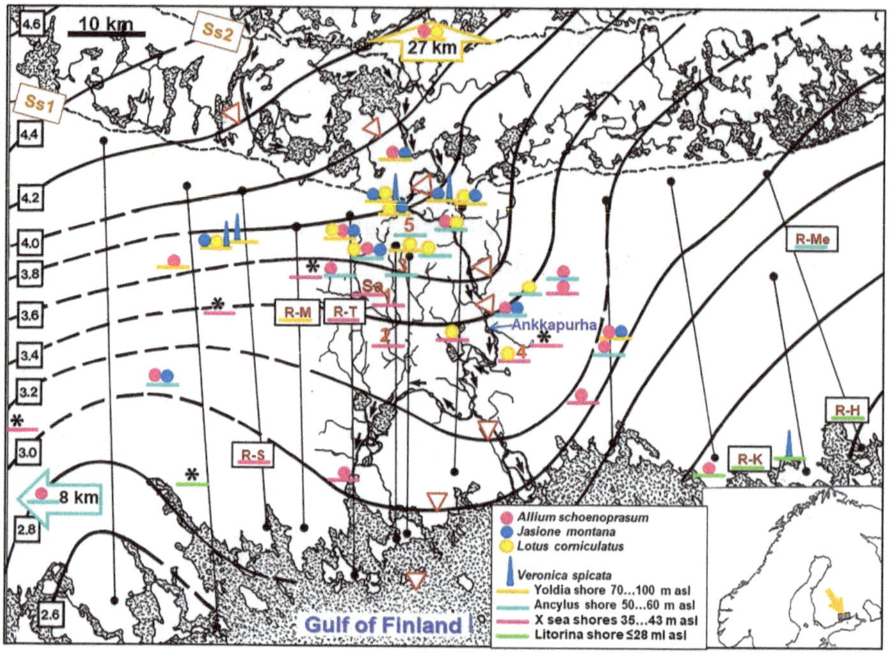

Fig. 1.1 Map showing the loci of sampling, four relict littoral plant species in inland sites and observations with their altitudes, land upheaval isopleths, the Kymi river and some ridges. **1** = Kaurisselkä; **2** = Pakinkangas; **3** = Raavuori; **4** = Välikangas; **5** = Ahkoja area; **Sa** = the Saarela site. The inland occurrences of the original, littoral, and rare plant species, *Allium schoenoprasum*, *Jasione montana*, *Lotus corniculatus*, and *Veronica spicata*, on different ancient shore levels are illustrated by the colors of the bars. The short-dash broken lines show the sand-moraine end-ridges, **Ss1** = the First Salpausselkä, and **Ss2** = the Second Salpausselkä. South of Ss1, only the waterways belonging to the drainage basin of the Kymi River were drawn. The rock art sites in frames: **R-H** = Himankirkko, Virolahti; **R-K** = Kummikirkonkivet, Virolahti, see Fig. 4.1a; **R-M** = Mukkokallio, Lapinjärvi; **R-Me** = Metsäisenjärvi, Miehikkälä; **R-S** = Svangan, Liljendal (Loviisa); **R-T** = Tyynelä, Elimäki (Kouvola), see Figs. 4.1b and 4.2. The five asterisks show the evident sites of inland ship wrecks, or parts of wrecks, as permitted to allocate from the information sources (Hipping 1817; Reinholm 1839–; Pakkala 1902; details in Ahokas 2003a). The 12 line segments connect the ends of the traceable glaciofluvial sand ridges with recognized clay junctions. The tilting of the clay contact levels in the four lines touching the Kymi is 1.08 ± 0.04 m/km (mean ± SE), and in the other lines, 1.48 ± 0.05 m/km; $t = 4.916$, $df = 10$, $p = 0.0006$ (Ahokas 2003a). Annual land upheaval rates in millimeters in the frames to the left and their isopleths according to Kääriäinen (1963). The red triangles point to the maxima of the regional upheaval isopleths, which are roughly followed by the route of the Kymi (black arrows) southwards of Ss2. Ankkapurha is one of the rapids. The age estimates of the prehistoric shores shown with the descriptor colors: Yellow for the Yoldia Sea, *c.* 11.2–10.2 cal ka BP. Turquoise for the Ancylus Lake since *c.* 10.2 cal ka BP. Green for the Litorina Sea *c.* 8.2–6.3 cal ka BP. Violet for the X sea stages after the Ancylus Lake. The lower right insert shows the situation of the region in Fennoscandia (yellow arrow)

Although sea levels changed relatively quickly in the Baltic Basin, the sea qualities, salt levels, and characteristic biota did not change uniformly across different regions of the Baltic. The variable rates of land upheaval or sinking formed a source of variation in the altitudes of the events between sites. The major factors determining Holocene shore levels in the Baltic Basin were the occurrences and positions of continental ice, the level of water in the North Sea, regional and central Sweden land elevation changes, and the straits of Denmark through isostatic readjustment (Tikkanen and Oksanen 2002). The fall of the Baltic Ice Lake water level in South Finland was dated to 11.4 to 11.2 ka ago (Hyttinen et al. 2014). The Yoldia Sea followed the Baltic Ice Lake and ended in the less salty Ancylus Lake, which was dated to have begun in Central Finland *c.* 10.20 to 10.15 cal ka BP (Ojala et al. 2005). The Litorina Sea transgression has been dated between 8.2 and 6.3 cal ka BP on the SE coast of Finland (Miettinen 2002). Within these major high stands, shore altitudes are complicated by other minor transgressions and regressions which may have formed newer shore levels above the altitudes of the older ones (Jantunen and Donner 1996). Simultaneously with the eustatic sea level changes, the nonuniform land upheaval rates inclined the whole Finland-to-be towards southeast and caused a valley anomaly in the region where the Kymi river took its course (Fig. 1.1). In the zone between the ancient shores of the Ancylus Lake and the Litorina Sea, there are up to nine raised beach levels south of the Ss1.

The development of the Kymi river south of the Ss1 has been little studied. The Ankkapurha (Fig. 1.1) is the second last of the rapids downstream in the Kymi river before it branches to its delta (Blomqvist 1911; Halme 1969). In 1911, the industry at the Ankkapurha rapids utilized 75 m^3/s of water falling 5 and 6.5 m to drive various machinery (Blomqvist 1911). However, behind the Ankkapurha name, there seems to be a modest meaning, 'Duck-brook', in the modern Finnish spelling *Ankkapuro* (Pakkala 1902), Anckapora (Teitti 1555–1556), Anckaborra (Bergström 1703), and in its derivatives written by Jönström (1774–1775), Ankapuro Forss (Swedish *fors* = 'rapids') and the Finnish Ankapuronnotko meaning 'Duck-brook's depression'. The current water level above the rapids (built in 1869–1872) is 30 m asl, and below, 22 m asl in the rapids bowl, where the eroded, deepest point is 13 m below the water surface (MapSite 2022–2024).

Saarnisto (2004) thought that the Ankkapurha rapids had developed during the 3500 years between 9.65 and 6.15 cal ka BP without showing physical dating. The earliest age limit is likely too old, since the land currently lies lower than 40 m above sea level (asl) around Ankkapurha, which was under the Ancylus Lake. Then, there occurred no Kymi river in the modern sense with its rapids.

The Holocene discharge of the proto-Kymi through Ss1, forming the Keltti rapids, 27 km NNW of the Ankkapurha, is dated from the bifurcated sands south of the Keltti rapids (Fig. 1.2). Here, their sand is typically less than 1 m thick, sitting on the Weichselian clay plain and, it apparently originated from the Ropakonmäki esker. The Ropakonmäki is *c.* 1 km downstream of the Keltti rapids, and up to 1.95 km north of the sands (Rainio et al. 1989a, b). The sands are found in two parts. The western part is *c.* 6.8 ha, and the eastern part, *c.* 7.2 ha (Rainio et al. 1989a). The sand-free intervening strip between them (Fig. 1.2) is interpreted as indicating proto-Kymi flow as the strip projects towards the apparent outburst channel in the Ropakonmäki esker. It is unrelated to the current Kymi River, which turns towards the southeast at the sands.

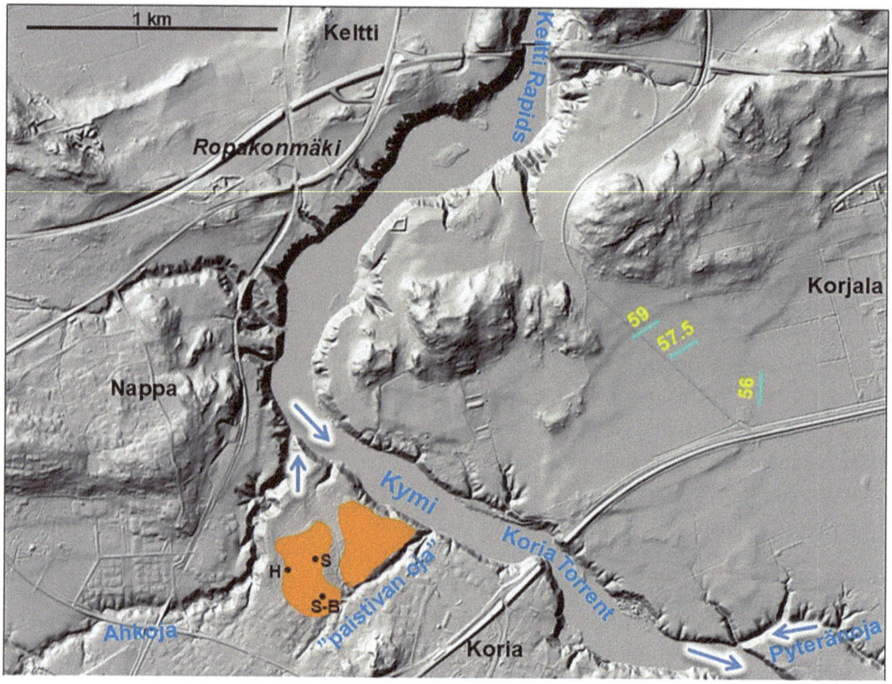

Fig. 1.2 Hillshade view of the Ahkoja area with the bifurcated Koria sand plains (marked with brown) on the clay plain. The water has discharged the First Salpausselkä ridge in Keltti and Ropakonmäki, from where at least part of the sands were washed out. The sites of the dated samples: **S-B** = Shfd16169, Beta-469432 and Beta-688480; **H** = Hv-24190; **S** = Su-3337. The junction angles of the Ahkoja and Pyteränoja tributaries "flowing against" the Kymi stream suggest that they were originally branches of the proto-Kymi and flowed then in the opposite direction. There are significantly more gorges on the proximal sides of Ahkoja and Pyteränoja (Ahokas 2003a). The "paistivan oja" name prior to 1788 (Oksanen 1985: p. 198) described frequent floodings (*paisuttaa, paisua*). The rocky, narrow passage of the Koria Torrent has been apt to be blocked by drifting ice and freeze to the bottom. The three ancient shore banks of the Ancylus Lake are detectable at 59, 57.5 and 56 m asl underlined ( ) on the clay to the east. Hillshade map from MapSite (2022–2024)

1.2 Preface to Historical Floods of the Kymi River

About half a kilometer after the sands, the river narrows to Korianvirta, "Koria Torrent", in a shallow, rocky passage (Fig. 1.2). Koria Torrent has been a site of ice blockages, which in turn could cause flooding. The ice blockages, sometimes mediated by the supercooling of water in ancient people's unmodified, natural rivers, with their flooding effects, were feared. According to oral tradition, the ice blockage was raised among the gods by the ancient Finns, known as Hyytö or Hyytämöinen (Ganander 1789). *Hyytö* is now a Finnish hydrological term for ice blockage (Kokkonen 1926). Ice blockages in the Kymi rapids have been observed in modern times (Blomqvist 1911; Kokkonen 1926; Ulvinen 1960). The *hyytö* blockage may begin when supercooled water forms ice on the bottom rocks in shallow rapids (Seppovaara 1988: p. 71; Hemmilä 2017) and can cause a rapidly rising flood (Alfthan 1908: p. 50). Additionally, an increased number of cobbles can move with the river current after the supercooled water freezes on them (Hemmilä 2017), as this decreases their apparent density.

Because of an ice blockage at Koria Torrent, major floods may have occurred during the Little Ice Age, at around 350 BP, when fluvial sand accumulated in the join of the Pärnäjoki and Kukonoja rivulets (*c.* 60°49′ N, 26°32′ E; Rainio et al. 1989a), which received flooding water from the Kymi (see below). Fine wood charcoal particles within fluvial sand were dated (Pärnäjoki-1, Hel-4438), giving a radiocarbon age of 553–294 cal a BP (Ahokas 2003a). The southward flooding from the Kymi via Ahkoja rivulet (Fig. 1.2) and subsequent lagoons towards the Pärnäjoki may have been regular still in the Middle Ages, making the Pärnäjoki an important fishing river, probably for spawning salmonids, about which fishing courts made decisions in 1460 and 1461 (Ahokas 2003a). Floods in Kymi in the 1500s and 1600s destroyed or damaged fishing constructions in the river (Halme 1969) and on the other hand, illegal constructions caused blockages (Oksanen 1981). Koria, a name derived from Korja, evidently has the ancient meaning of 'swampy river' (Ahokas 2003a); note also the toponym Korjala on the eastern side (Fig. 1.2). The Koria sand plains occur in Vilppula village of the centuries-old municipality Elimäki, annexed to the Kouvola town in 2009. The recorded oral tradition of local people tells that the territory of the Elimäki municipality (parish) has been 'part of the great bottom of the Kymi' (Pakkala 1902). Measures to control the effects of the floods in Elimäki were sought in the applications since 1826 (Blomqvist 1911). Local people also pointed out ancient or old flood channels to Reinholm (1839–). Many ancient toponyms had the Finnish *-saari*, *saare-*, or *-saarento* part ('island'), and three *napa* toponyms ('hill-top', 'elevation above flood') that remind of times when these elevations used to stay above the floods (Ahokas 2003a).

After the Keltti rapids, the proto-Kymi streamed southwards, eroding the branching rivulet called Ahkoja, now a tributary (Fig. 1.2). Flooding water flew via the Ahkoja into lagoons, later appearing as the swamp area comprising Martinaro (*c.* 60°51′ N, 26°33′ E), Karjusuo (*c.* 60°50′ N, 26°34′ E) and Hevossuo (*c.* 60°50′ N, 26°32′ E). The water streamed from them as the Pärnäjoki river southwards ending in the Elimäenjärvi lake, which in the 1600th century was called Ilmajärvi; *ilma-* is an ancient Finnish geographic appellative for a watercourse apt to flooding (Ahokas 2003a; for discussion see the Sect. 4.6). The Karjusuo swamp was apparently maintained wet by floods; people who had sunk water-wells in Karjusuo in modern times told that its soil has remains of rush (Punttila 1960) as macroscopic subfossils which also occur under clay sediment on the N side of the Kymi in this area (Hemmilä 2018: p. 98) *c.* 50 m asl. The "paistivan oja" gorge bordering the SE sides of the Koria sand plains (Fig. 1.2) was known on a map before 1788 (Oksanen 1985: p. 198). The name is derived from the Finnish verb *paisuttaa* (<*paisua*), of the archaic meaning 'to inundate', and the Finnish *aro* part in its more modern name Näsinaro also reminds of a wetland. In the Martinaro, then apparently still a quagmire, the people of the nearby Nappa village looked for shelter against the torching and marauding Russian military groups during the wars in the eighteenth century, while the people of the Vilppula village looked for shelter in Mannari (*c.* 60°48′ N, 26°32′ E; Pakkala 1902) being on a hill largely surrounded by the past wetlands. Enlargement of the Ankkapurha rapids (Fig. 1.1) has occurred in a short historical time. The soil erosion on the shore of the rapids bowl was recorded and measured by the land surveyor H. Richenberg (1727). He illustrated this process with a map and an account that in 20 years' time, 12 *tunnorLd*, or 5.9 ha of the earlier ploughed land had disappeared with the river from Inkeroinen, i.e., the NE side of the rapids bowl. Richenberg also allocated on his map and accounted for an existing rock in the eroded field. According to other records, there was in the rapids bowl a rock called Arranparantaja, 'Plough-improver', known in the early nineteenth century to have been in the field formerly. The streaming water eroded the field soil away, and the rock situated 50 fathoms (*c.* 90 m) from the shore in about 1870 (Reinholm 1839–). The name was explained from a collision with a plough against the Arranparantaja, which had reformulated the tool, apparently improving its performance. Arranparantaja still appeared in the map drawn in about 1870 by P. K. Wulisma, who was the owner of the Inkeroinen farm on the NE and E side of the rapids from 1865 to 1871 (Häkkinen and Sippu 2000). Wulisma's archival map was published in Ahokas (2003b) and Schulz (2004).

References

Ahokas H (2003a) Major discharge and floods of the Kymi river (Kymijoki) over the Salpausselkä I esker to coastal South Finland traced by ^{14}C datings, unequal land upheaval, oral tradition and onomastics. Kave, Helsinki

Ahokas H (2003b) Novgorodilaisten vuoden 1311 terroriretken reitistä. Kymenlaakson museotiedote 2003(1–2):25–36

Ahokas H (2018) Ruoholaukan uusi löytöpaikka muinaisrannalla Kymenlaaksossa ja mustan biofilmin merkitys. Summary: Allium schoenoprasum on the prehistoric lake shore at a recently discovered inland site in SE Finland. Lutukka 34:104–111. https://tietopankki.luomus.fi/lehdet-ja-kirjat/lutukka/

Alfthan M (1908) Bidrag till kännedom af Kymmene älf. Résumé: Contribution à l'étude du Kymmene. Fennia 27(3):1–96, appendices

Ancient shorelines (2013) Special data products. The Geological Survey of Finland, Espoo

Bergström EA (1703) Geometrisk delineation öfwer Umeliocki Aniala och Wärälä bÿÿars skogsegor med deras twistigheteer whilka finnes belägne uti Kymmenegårdz lähn och Elimä sochn råernas beskrifning kring alla 3 bÿars ägor. Anjala G138, 2/1–2. The National Archives of Finland, Helsinki. (Archival map)

Björk S (1967) Ecological investigations of Phragmites communis. Fol Limnol Scand 14:1–248, Tables

Blomqvist E (1911) Kymijoki ja sen vesistö. Lisiä Suomen hydrografiaan, vol II(1). Hydrografinen toimisto, Helsinki, pp 1–235

Boyes BM, Pearce DM, Linch LD, Nash DJ (2024) Younger Dryas and early Holocene ice-margin dynamics in Northwest Russia. Boreas 53:376–400. https://doi.org/10.1111/bor.12653

Ganander C (1789) Mythologia fennica. Christfrid Ganander and Frenckellska Boktryckeriet, Åbo. The 4th facsimile print (1984). Suomalaisen Kirjallisuuden Seura, Helsinki

GTK maankamara (2024). https://gtkdata.gtk.fi/maankamara

Häkkinen V, Sippu S (2000) Mämmälän kylän talot ja suvut. V Häkkinen, Helsinki

Halme E (1969) Kyminkartanon läänin kalastus 1500- ja 1600-luvulla. MSc thesis, University of Helsinki

Hemmilä R (2017) Pommi pohjassa. Koala-Kustannus, Helsinki

Hemmilä R (2018) Keltist Mylsään. R Hemmilä, Koria

Hipping AJ (1817) Beskrifning öfwer Perno socken i Finland. MC Iversen, St. Petersburg

Hyttinen O, Eskola KO, Kaakinen A, Salonen V-P (2014) First direct age determination for the Baltic Ice Lake/Yoldia Sea transition in Finland. GFF 136:398–405. https://doi.org/10.1080/11035897.2013.813581

Jantunen T, Donner J (1996) The formation of raised beaches in southern Finland during the Ancylus and Litorina stages. Bull Geol Soc Finl 68(2):34–39. sgs_bt_068_2_pages_034_039.pdf

Jönström T (1774–1775) Tredje charta delen öfwer Mämmälä samfällighets ägor uti Weckelax socken ock Kymmene härad af Wiborgs län. Afmåtte till skattläggning. Mämmälä G18, 9/1-9. The National Archives of Finland, Helsinki. (Archival map)

Kääriäinen E (1963) Land uplift in Finland computed by the aid of precise levellings. Fennia 89(1):15–19, map

Kokkonen P (1926) Kymijoen perkauksesta johtuva maanarvon nousu. Referat: Die Steigerung des Bodenwertes infolge der Reinigung des Kymijokiflusses. Suomen Maataloustiet Seur Julk 11:1–46

Kutschera L (1960) Wurzelatlas mitteleuropäischer Ackerunkräuter und Kulturpflanzen. DLG-Verlag-GmbH, Frankfurt am Main

MapSite (2022–2024) The National Land Survey of Finland, Helsinki. maanmittauslaitos.fi/en/e-services/mapsite

Miettinen A (2002) Relative sea level changes in the eastern part of the Gulf of Finland during the last 8000 years. Ann Acad Sci Fenn Geol–Geogr 162:1–100

Ojala AEK, Heinsalu A, Saarnisto M, Tiljander M (2005) Annually laminated sediments date the drainage of the Ancylus Lake and early Holocene shoreline displacement in central Finland. Quat Int 130:63–73

Ojala AEK, Palmu J-P, Åberg A, Åberg S, Virkki H (2013) Development of an ancient shoreline database to reconstruct the Litorina Sea maximum extension and the highest shoreline of the Baltic Sea basin in Finland. Bull Geol Soc Finl 85:127–144. https://doi.org/10.17741/bgsf/85.2.002

Oksanen E-L (1981) Anjalan historia. Anjalan kunta, Anjalankoski, pp 13–14, 29–691

Oksanen E-L (1985) Elimäen historia. Elimäen kunta, Elimäki

Pakkala EW (1902) Luettelo Pernajan kihlakunnan suomalaisten pitäjien muinaismuistoista. Suomen Muinaismuistoyhdistyksen pöytäkirjat. 10/X 1902. www.kyppi.fi/to.aspx?id=129.146408. (Manuscript)

Punttila M (1960) Karjusuo and Mannari, Elimäki. Names Archive. Institute for the Languages of Finland, Helsinki. (Archival record)

Rainio H, Mäkilä M, Grundström A, Paukola T, Urvas L (1989a) Koria 3113 04. Maaperäkartta 1:20 000. Geologian tutkimuskeskus, Maatalouden tutkimuskeskus and Maanmittaushallitus, Helsinki. (Soil map)

Rainio H, Mäkilä M, Grundström A, Urvas L (1989b) Kuusankoski 3113 05. Maaperäkartta 1:20 000. Geologian tutkimuskeskus, Maatalouden tutkimuskeskus and Maanmittaushallitus, Helsinki. (Soil map)

Rämö OT, Turkki V, Mänttäri I, Heinonen A, Larjamo K, Lahaye Y (2014) Age and isotopic fingerprints of some plutonic rocks in the Wiborg rapakivi granite batholith with special reference to the dark wiborgite of the Ristisaari Island. Bull Geol Soc Finl 86:71–91. https://doi.org/10.17741/bgsf/86.2.002

Reinholm HA (1839) H. A. Reinholmin arkisto. Suomen Muinaismuistoyhdistys. Archives of the National Board of Antiquities, Helsinki. (Manuscripts)

Richenberg H (1727) Geometrisk charta öfwer Mämälä bys åker belägit i Kymmenegårdz lähn och Wäckelax sochn afmått anno 1727. Mämmälä G17, 5/1. The National Archives of Finland, Helsinki. (Archival map)

Saarnisto M (2004) Muinainen Kymijoki. In: Uino P (ed) Ammoin Ankkapurhassa. Museovirasto, Helsinki and Stora Enso, Anjalankoski, pp 38–41. Book summary: Ancient Ankkapurha. Research at Ankkapurha, pp 76–79

Sahala L (1988) Korian ympäristön saviselvitys. Geol Surv Finl Rep P3.31.4.003. https://tupa.gtk.fi/raportti/arkisto/p31_4_003.pdf

Schulz E-L (2004) Ankkapurhan tutkimushistoriaa. In: Uino P (ed) Ammoin Ankkapurhassa. Museovirasto, Helsinki and Stora Enso, Anjalankoski, pp 11–21. Book summary: Ancient Ankkapurha. Research at Ankkapurha, pp 76–79

Seppovaara O (1988) Kymijoki. Kouvolan Kirjapaino Oy, Kuusankoski

Simonen A, Laitala M (1970) Geological Map of Finland 1:100 000. Pre-Quaternary rocks. Sheet 3023 + 3014—Kotka. Geological Survey. https://tupa.gtk.fi/kartta/kallioperakartta100/kp_3023_3014.pdf

Simonen A, Lehijärvi M (1963) Geological Map of Finland 1:100 000. Pre-Quaternary rocks. Sheet 3113—Kouvola. Geological Survey. https://tupa.gtk.fi/kartta/kallioperakartta100/kp_3113.pdf

Teitti J (1555–1556) Jakob Teitts klagomålsregister emot adeln i Finland. Microfilms 215j and 215k. The National Archives of Finland, Helsinki. (Manuscript)

Tikkanen M, Oksanen J (2002) Late Weichselian and Holocene shore displacement history of the Baltic Sea in Finland. Fennia 180:9–20. CD. https://fennia.journal.fi/article/download/3760/3551

Tynni R, Hyyppä J, Valovirta V (1976) Lapinjärven kartta-alueen maaperä. Summary: quaternary deposits in the Lapinjärvi map-sheet area. Suomen geologinen kartta 1:100 000. Maaperäkartan selitykset, 3022 Lapinjärvi. Geologinen tutkimuslaitos, Espoo

Ukkonen P, Mannermaa K (2017) Museoviraston julkaisuja, vol 8, pp 1–240

Ukkonen P, Lunkka JP, Jungner H, Donner J (1999) New radiocarbon dates from Finnish mammoths indicating large ice-free areas in Fennoscandia during the Middle Weichselian. J Quat Sci 14:711–714. https://doi.org/10.1002/(SICI)1099-1417(199912)14:7<711::AIDJQS506>3.0.CO;2-E

Ulvinen A (1960) Luonto. In: Niinisalo KV (ed) Kymin historia I. Kymin seurakunta and kunta, Kymi, pp 9–29

Valovirta V (1972) Kymen läänin rannikkoalueen maaperä. Summary: Quaternary deposits in the coastal area of the Kymi district. Suomen geologinen kartta 1:100 000. Maaperäkartan selitykset, 3023 + 3012 Kotka, 3024 Karhula, 3041 + 3043 Haapasaari, 3042 Hamina. Geologinen tutkimuslaitos, Espoo

Chapter 2
Materials and Methods

Abstract Inland occurrences of coastal plant species were studied. Their sources were common museal herbaria, Kasviatlas scientific plant database, recently collected specimens and published field observations in the region with known altitudes related to inland paleoshorelines. Soil samples protected from light were collected for chronology dating with luminescence. Soil samples for radiocarbon dating, grain size determination plus organic and charred material were collected to date and to outline the development during the Holocene. The principles of the luminescence dating used are presented. The major sources of records and memorials concerning the ancient shores, channels, floods and human life were presented. These included the H. A. Reinholm collection he began in 1839, and the memories from the region collected with a scholarship by E. W. Pakkala and deposited in 1902; Reinholm and Pakkala were academic people.

Keywords Allium schoenoprasum · Herbaria · Jasione montana · Kasviatlas plant database · Kouvola · Lotus corniculatus · Luminescence dating · Sand grain sizing · Radiocarbon dating · Veronica spicata

Insight into the timing and paleoenvironment was obtained through proxy methods, including littoral plant species distributions, shipwrecks, and rock art, as well as absolute chronology methods, such as luminescence and radiocarbon dating. Insights were further drawn from oral traditions.

H. Ahokas et al., *Paleoshorelines and Their Relict Littoral Plants*, SpringerBriefs in Earth Sciences, https://doi.org/10.1007/978-3-032-17723-0_2

2.1 Proxy Methods: Observations of Four Littoral Plant Species in Inland Habitats

Four rare littoral relict plant species occur in the inland and are thought to relate to past shores. With the exception of some modern apophytic occurrences of trefoil (*Lotus corniculatus*) and cultural remains of chive (*Allium schoenoprasum*) associated with the modern dwelling, the occurrences of trefoil, chive, sheep's-bit (*Jasione montana*) and spiked speedwell (*Veronica spicata*) (Figs. 3.1, 3.2, 3.3 and 3.4) could have established themselves on the littoral levels of the Yoldia Sea and now find themselves existing inland. These four species today are rare inland in Finland, with their habitats having been protected from anthropogenic disturbance due to their remoteness. The chive and trefoil are fairly common but only on the present shores of the Gulf of Finland and of its islands (Eklund 1958; Kujala and Ulvinen 1964). To establish inland occurrences of the four species, this study used herbarium samples, botanists' published reports or records (Kujala and Ulvinen 1964; Taarna 1968, 2000, 2007; T. Rintanen in Floristic Archives of the Botanical Museum H in Helsinki 1984; Haikonen 1986; Horppila 2001; Ahokas 2006, 2018, and in Lampinen and Lahti 2019; A. Kiviniemi in Lampinen and Lahti 2019; A. Kiviniemi personal e-mails to H.A.). Occurrences were then geo-located with their altitudes established according to 1:20,000 topographic maps using editions published since 1980 based on surveys by the National Land Survey of Finland.

2.2 Proxy Methods: Landmarks, Records and Memorials Concerning the Ancient Shores and Streams

The position and extent of ancient shorelines and their levels in the region were compiled from published records (Hyyppä 1937; Miettinen 2002; Tikkanen and Oksanen 2002; Pesonen 2003; Ancient shorelines 2013; Ojala et al. 2013). This was supplemented on sandy areas, using hillshade maps, which sometimes show the positions of the raised beaches (MapSite 2022–2024). Information on shipwreck localities and details of ancient river channels and sea shorelines were also derived from archival collections of oral traditions (Hipping 1817; Reinholm 1839–; Pakkala 1902). Historical data concerning rivers, their floods, rapids and water levels affecting human life were derived from Teitti (1555–1556), Bergström (1703), Richenberg (1727), Ganander (1789), Reinholm (1839–), Blomqvist (1911), Kokkonen (1926), Halme (1969) and Oksanen (1981). Altitudes of rock art have often been found to relate to ancient sea, lake and river water levels and as such, rock art has been found in the region south of Ss1 (Kivikäs 1995; Miettinen 2000; Ahokas 2012, 2015) and was included in this study.

2.3 Chronology Sample Collection

The luminescence samples aimed to determine the spread of sand on the glacial clay plain (Koria sand plains), the formation of the major raised beach of the Raavuori rock, and glaciofluvial sand directly under selected raised beaches (Kaurisselkä and Pakinkangas). Additionally, the sampling looked to date the glaciofluvial sand deposit of the Välikangas esker, which later formed raised beaches relating to sea shoreline levels. The samples of sand or gravel for the luminescence dating were taken from vertical exposures created by hand digging or by cleaning back about 10 cm of existing machine-made cuts. Steel tubes with a diameter of 5 cm and a length of 35 cm were horizontally placed into exposures to collect sediment that was unexposed to sunlight in them. A black polypropylene foam stop covered the free end of the tube from the beginning and the ends were sealed with black plastic stops immediately after withdrawal.

As the project spanned several years, luminescence dating was performed at two facilities. Samples collected in 2016 were dated at the luminescence facility of the University of Sheffield (Shfd), and samples collected in 2019 were dated at the Netherlands Centre for Luminescence Dating (NCL). The specific methods used at each facility are detailed below and in Chap. 6. The samples for radiocarbon determinations were also collected to date floodings or splashing over the beach banks.

2.4 Luminescence Dating in Sheffield: Shfd

The sample measured at Sheffield (Shfd16169) was prepared under red light conditions following the procedure to extract and clean quartz in the size range of 90–250 μm as outlined in Bateman and Catt (1996). Measurements at the single aliquot level used a Risø TL DA-20 luminescence reader with the sample prepared onto aliquots comprising a *c.* 9 mm diameter quartz grain monolayer on 9.6 mm diameter stainless steel discs. Stimulation was provided by blue/green LEDs, and luminescence detection was through a Hoya U-340 filter. To establish the samples' palaeodose (D_e), the single-aliquot regenerative (SAR) approach of Murray and Wintle (2003) was applied. The most appropriate preheat temperature for the sample was selected using a dose recovery preheat plateau test, which resulted in the selection of a preheat temperature of 200 °C for 10 s. Feldspar contamination was found as a consequence of which an IR (infra-red) wash was inserted into the SAR protocol prior to each OSL (optical stimulated luminescence) measurement. All aliquots where the ratio of first and last dose point (recycling ratio) as measured in SAR exceeded ±10% of unity were excluded from further analysis. Measurements showed the sample had moderately good

OSL characteristics with some contribution from an unstable medium OSL component, indicated by only moderate depletion with stimulation time (Fig. 6.1a). However, the SAR growth curve data were well fitted with an exponential function, aliquots had good SAR recycling and a dose recovery test was able to recover a known dose. No aliquots were saturated, which would have indicated a sediment whose antiquity is beyond the limits of the SAR OSL technique. Multiple replicates were measured to better understand sample D_e variability. This showed the D_e data distribution to be broadly normally distributed and reproducible, and once outliers were excluded, the overdispersion (OD) was within the range of what would be expected for a well-bleached sand. The D_e value derived for age calculation purposes was therefore used, based on the central Age Model of Galbraith et al. (1999). Dose rates for this sample were based on inductively coupled plasma mass spectroscopy measurements of potassium (K), thorium (Th), rubidium (Rb), and uranium (U). Elemental concentrations were converted to annual dose rates using data from Aitken (1998), Adamiec and Aitken (1998), and Marsh et al. (2002). Attenuation of dose by grain size followed Mejdahl (1979), and attenuation by moisture, assuming present-day values were representative, with a ±5% error to incorporate seasonal and longer-term fluctuations in moisture that the sample may have endured since burial. The contribution to dose rates from cosmic sources was calculated using the expression published in Prescott and Hutton (1994). The Sheffield age (Shfd16169) is quoted in years from the present day (2017) with one sigma confidence intervals which incorporate systematic uncertainties with the dosimetry data, uncertainties with the palaeomoisture content, and errors associated with the D_e determination.

2.5 Luminescence Dating in Wageningen: NCL

Preparation and measurement of five samples were performed under amber light conditions at the Netherlands Centre for Luminescence Dating. In light of the data from the initial sample dated at Sheffield and as a result of further tests, which confirmed poor sensitivity of the quartz OSL signal, further dating at NCL was performed on feldspar minerals at the single-grain level. Grains of 212–250 μm were retrieved by wet sieving and treated with HCl and H_2O_2 to remove carbonates and organics. Magnetic grains were removed with a Frantz Magnetic Barrier Laboratory Separator Model LB-1 (Porat 2006), and the non-magnetic fraction was density separated to obtain grains lighter than 2.58 g/cm^3, thereby targeting the potassium-rich feldspar grains. No hydrofluoric acid etching was applied to the feldspar grains.

Luminescence measurements were performed on single grains using a Risø TL/OSL reader equipped with a 140 mW 830 nm IR laser and a Schott RG 780 filter, placed directly in front of the laser and using single-grain discs with holes 300 μm in diameter (Bøtter-Jensen et al. 2003). A post-infrared infrared stimulated luminescence (pIRIR) protocol inspired by Reimann et al. (2017) and Brill et al. (2018), detailed in Chap. 6 (Table 6.2), was used to obtain the pIRIR-225 signal for dating. The pIRIR-225 stimulation time was 1.67 s; the signal was integrated over the first 0.10 s after the IR laser was switched on, and the background was integrated over the last 0.37 s. The pIRIR-225 yielded a bright signal (Fig. 6.4), and grains were accepted if recycling was within 10% and recuperation was below 10% including uncertainties. The validity of the measurement protocol was assessed using a dose recovery test performed on three samples for which sufficient material was available. Results were somewhat below unity, yielding ratios of 0.84 ± 0.02 (NCL-7320041; n = 136), 0.81 ± 0.02 (NCL-7320042; n = 105) and 0.89 ± 0.02 (NCL-7320043; n = 117) for the samples. These values are below unity, which may suggest that the D_e values underestimate the burial dose by *c*. 10%. We opted not to apply a correction because the best treatment for this potential underestimate is not yet known. A fading test (Auclair et al. 2003) performed on three multigrain aliquots showed negligible fading of the pIRIR-225 signal (g-value = 0.95 ± 0.07% per decade, see Fig. 6.5), so a fading correction was not applied. The D_e as estimated using an iterated mean because the samples appeared sufficiently bleached. For samples where all grains were in saturation, we used the sample-average 2D0 value as a minimum D_e value.

Dose rate was determined from radionuclide activities measured on a gamma spectrometer for bulk surrounding sediment ground and prepared within a wax puck. Following Smedley et al. (2012), we assumed a K-content of 10 ± 2%. Cosmogenic contributions were determined following Prescott and Hutton (1994), assuming instant burial, grain-size attenuation was determined following Mejdahl (1979), and attenuation by water content and organics was determined following Aitken (1998). We used the *in situ* water content for samples with water content greater than 5%, and assumed a water content of 5% for samples with *in situ* water contents below this value. All water contents are expressed in percent dry weight with 25% relative uncertainty. We used the *in situ* organic contents for each sample with a 10% relative uncertainty. NCL luminescence ages are reported in ka relative to 2022, the year of dating.

2.6 Radiocarbon Dating and Calibration

Carbonaceous grains were individually picked from the soil under a stereomicroscope, cleaned of unwanted contaminants, and washed in water. The primary radiocarbon ages of the samples coded with Hel- (Jungner and Sonninen 1998), Hv- (Jelen and Geyh 1986), and Su- (Mäntynen et al. 1987; Äikää et al. 1992) were corrected with $\delta^{13}C$ values. The Beta samples were acid/alkali/acid treated and analyzed with AMS using international standards and corrected with $\delta^{13}C$ values (Beta Analytic 2018). Calibration of radiocarbon ages for this and those previously published elsewhere was done with the OxCal4.4 online program (OxCal–ORAU 2022–2024). The calibrated ages have a 95.4% probability, if not otherwise stated. The ground radioactivity of U and Th may increase the ^{14}C content in nitrogen-containing material. The data of the aeroradiometric measurements mapped in 50 × 50 m pixels were purchased from the Geological Survey of Finland. Some earlier samples of radiocarbon dating referred to here originate from the years between 1998 and 2001 and are indicated with reference to Ahokas (2003) in Table 3.2.

2.7 Sand Grain Size Determination

Sand lots of 150 g from dug, representing room-dry samples of kilograms, were fractionated in a set of six sieves and a pan (Ø 20 cm) of stainless steel (Endecotts, QAQC, VA, USA) with the openings of 6.3, 2.8, 2, 0.6, 0.063, and 0.02 mm. Each sieving lasted 5 min with rapid circular motions with hands for about 3 s and rapping the sieve set against the bench after each 3-s motion. The recovery of the samples was $\geq$99.4%.

References

Adamiec G, Aitken MJ (1998) Dose-rate conversion factors: update. Ancient TL 16:37–50. ancienttl.org/ATL_16-2_1998/ATL_16-2_Adamiec_p37-50.pdf

Ahokas H (2003) Major discharge and floods of the Kymi river (Kymijoki) over the Salpausselkä I esker to coastal South Finland traced by ^{14}C datings, unequal land upheaval, oral tradition and onomastics. Kave, Helsinki

Ahokas H (2006) Tähkätädykkeen, ruoholaukan ja keltamaitteen reliktiesiintymistä sisämaan muinaisrannoilla. Summary: inland occurrences of Veronica spicata, Allium schoenoprasum and Lotus corniculatus at ancient Baltic Sea shore levels. Lutukka 22:45–57

Ahokas H (2012) Crop evolution under fire. Interdiscip Biol Agric Linguist Antiq 5:1–194

Ahokas H (2015) Aikaisemmin tunnistamattomia muinaistaidekuvia, hieroglyfi, petroglyfejä ja kalliomaalauksia Kaakkois-Suomesta. Abstract: Recently found hieroglyph, other petroglyphs and rock paintings in South-East Finland and about the helsink appellative in toponyms. Interdiscip Biol Agric Linguist Antiq 6:1–63

Ahokas H (2018) Ruoholaukan uusi löytöpaikka muinaisrannalla Kymenlaaksossa ja mustan biofilmin merkitys. Summary: Allium schoenoprasum on the prehistoric lake shore at a recently discovered inland site in SE Finland. Lutukka 34:104–111. https://tietopankki.luomus.fi/lehdet-ja-kirjat/lutukka/

Äikää O, Mäntynen P, Kankainen T (1992) High-performance ^{14}C gas-proportional counting system applying pulse-shape discrimination. Radiocarbon 43:414–419

Aitken MJ (1998) An introduction to optical dating. Oxford University Press, Oxford

Ancient shorelines (2013) Special data products. The Geological Survey of Finland, Espoo

Auclair M, Lamothe M, Huot S (2003) Measurement of anomalous fading for feldspar IRSL using SAR. Radiat Meas 37:487–492. https://doi.org/10.1016/S1350-4487(03)00018-0

Bateman MD, Catt JA (1996) An absolute chronology for the raised beach and associated deposits at Sewerby, East Yorkshire, England. J Quat Sci 11:389–395. https://doi.org/10.1002/(SICI)1099-1417(199609/10)11:5<389::AID-JQS260>3.0.CO;2-K

Bergström EA (1703) Geometrisk delineation öfwer Umeliocki Aniala och Wärälä bÿÿars skogsegor med deras twistigheteer whilka finnes belägne uti Kymmenegårdz lähn och Elimä sochn råernas beskrifning kring alla 3 bÿars ägor. Anjala G138, 2/1–2. The National Archives of Finland, Helsinki. (Archival map)

Beta Analytic, Inc (2018) Introduction to radiocarbon determination by the accelerator mass spectrometry method. https://www.radiocarbon.cn/uploads/2018/07.pdf

Blomqvist E (1911) Kymijoki ja sen vesistö. Lisiä Suomen hydrografiaan, vol II(1). Hydrografinen toimisto, Helsinki, pp 1–235

Bøtter-Jensen L, Andersen CE, Duller GAT, Murray AS (2003) Developments in radiation, stimulation and observation facilities in luminescence measurements. Radiat Meas 37:535–541. https://doi.org/10.1016/S1350-4487(03)00020-9

Brill D, Reimann T, Wallinga J, May SM, Engel M, Riedesel S, Brückner H (2018) Testing the accuracy of feldspar single grains to date late Holocene cyclone and tsunami deposits. Quat Geochronol 48:91–103. https://doi.org/10.1016/j.quageo.2018.09.001

Eklund O (1958) Die Gefässpflanzenflora beiderseits Skiftet im Schärenarchipel Südwestfinnlands. Bidrag till kännedom af Finlands natur och folk, vol 101, pp 1–324, maps

Galbraith RF, Roberts RG, Laslett GM, Yoshida H, Olley JM (1999) Optical dating of single and multiple grains of quartz from Jinmium rock shelter, northern Australia: Part I, experimental design and statistical models. Archaeometry 41:339–364. https://doi.org/10.1111/j.1475-4754.1999.tb00987.x

Ganander C (1789) Mythologia fennica. Christfrid Ganander and Frenckellska Boktryckeriet, Åbo. The 4th facsimile print (1984). Suomalaisen Kirjallisuuden Seura, Helsinki

Haikonen V (1986) Päijät-Hämeen uhanalaiset ja harvinaiset kasvit. Päijät-Hämeen seutukaavaliiton julkaisuja A, vol 24, pp 1–51, appendices

Halme E (1969) Kyminkartanon läänin kalastus 1500- ja 1600-luvulla. MSc thesis, University of Helsinki

Hipping AJ (1817) Beskrifning öfwer Perno socken i Finland. MC Iversen, St. Petersburg

Horppila P (2001) Tähkätädykettä etsimässä Artjärvellä. Lutukka 17:74–77

Hyyppä E (1937) Post-glacial changes of shore-line in South Finland. Bull Commis Géol Finl 120:1–225, 2 appendices

Jelen K, Geyh MA (1986) A low-cost miniature counter system for radiocarbon dating. Radiocarbon 28:578–585

Jungner H, Sonninen E (1998) Radiocarbon dates. Dating Laboratory, University of Helsinki. Report 5:1–91

Kivikäs P (1995) Kalliomaalaukset muinainen kuva-arkisto. Summary: Paintings on rock—an ancient picture archive. Atena Kustannus Oy, Jyväskylä, pp 1–324, 334–336

Kokkonen P (1926) Kymijoen perkauksesta johtuva maanarvon nousu. Referat: Die Steigerung des Bodenwertes infolge der Reinigung des Kymijokiflusses. Suomen Maataloustiet Seur Julk 11:1–46

Kujala V, Ulvinen A (1964) Floristische Untersuchungen in Ost-Kymenlaakso in Südfinnland. Ann Bot Soc Zool Bot Fenn Vanamo 35(2):1–215

Lampinen R, Lahti T (2019) Kasviatlas 2018. Helsingin Yliopisto, Luonnontieteellinen keskusmuseo, Helsinki. http://koivu.luomus.fi/kasviatlas. (Plant distribution maps and database)

Mäntynen P, Äikää O, Kankainen T, Kaihola L (1987) Application of pulse-shape discrimination to improve the precision of the carbon-14 gas-proportional-counting method. Int J Radiat Appl Instrum (Part A) Appl Radiat Isotopes 38:869–873

MapSite (2022–2024) The National Land Survey of Finland, Helsinki. maanmittauslaitos.fi/en/e-services/mapsite

Marsh RE, Prestwich WV, Rink WJ, Brennan BJ (2002) Monte Carlo determinations of the beta dose rate to tooth enamel. Radiat Meas 35:609–616. https://doi.org/10.1016/S1350-4487(02)00065-3

Mejdahl V (1979) Thermoluminescence dating: beta-dose attenuation in quartz grains. Archaeometry 21:61–72. https://doi.org/10.1111/j.1475-4754.1979.tb00241.x

Miettinen T (2000) Kymenlaakson kalliomaalaukset. Abstract: Rock paintings in Kymenlaakso. Kymenlaakson maakuntamuseon julkaisuja 27:1–163

Miettinen A (2002) Relative sea level changes in the eastern part of the Gulf of Finland during the last 8000 years. Ann Acad Sci Fenn Geol–Geogr 162:1–100

Murray AS, Wintle AG (2003) The single aliquot regenerative dose protocol: potential for improvements in reliability. Radiat Meas 37:377–381. https://doi.org/10.1016/S1350-4487(03)00053-2

Ojala AEK, Palmu J-P, Åberg A, Åberg S, Virkki H (2013) Development of an ancient shoreline database to reconstruct the Litorina Sea maximum extension and the highest shoreline of the Baltic Sea basin in Finland. Bull Geol Soc Finl 85:127–144. https://doi.org/10.17741/bgsf/85.2.002

Oksanen E-L (1981) Anjalan historia. Anjalan kunta, Anjalankoski, pp 13–14, 29–691

OxCal–ORAU (2022–2024) Radiocarbon calibration. https://c14.arch.ox.ac.uk/oxcal.html

Pakkala EW (1902) Luettelo Pernajan kihlakunnan suomalaisten pitäjien muinaismuistoista. Suomen Muinaismuistoyhdistyksen pöytäkirjat. 10/X 1902. www.kyppi.fi/to.aspx?id=129.146408. (Manuscript)

Pesonen P (2003) Esihistoria. In: Aalto S (ed) Pornaisten historia 1. Pornaisten kunta, Pornainen, pp 23–55, 299–301 and 309–315

Porat N (2006) Use of magnetic separation for purifying quartz for luminescence dating. Ancient TL 24:33–36. https://ancienttl.org/ALT_24-2.pdf

Prescott JR, Hutton JT (1994) Cosmic ray contributions to dose rates for luminescence and ESR dating: large depths and long-term time variations. Radiat Meas 23:497–500. https://doi.org/10.1016/1350-4487(94)90086-8

Reimann T, Román-Sánchez A, Vanwalleghem T, Wallinga J (2017) Getting a grip on soil reworking—Single-grain feldspar luminescence as a novel tool to quantify soil reworking rates. Quat Geochronol 42:1–14. https://doi.org/10.1016/j.quageo.2017.07.002

Reinholm HA (1839) H. A. Reinholmin arkisto. Suomen Muinaismuistoyhdistys. Archives of the National Board of Antiquities, Helsinki. (Manuscripts)

Richenberg H (1727) Geometrisk charta öfwer Mämälä bys åker belägit i Kymmenegårdz lähn och Wäckelax sochn afmått anno 1727. Mämmälä G17, 5/1. The National Archives of Finland, Helsinki. (Archival map)

Smedley RK, Duller GAT, Pearce NJG, Roberts HM (2012) Determining the K-content of single-grains of feldspar for luminescence dating. Radiat Meas 47:790–796. https://doi.org/10.1016/j.radmeas.2012.01.014

Taarna J (1968) Tutkimuksia kallioiden kasvistosta, kasvillisuudesta ja pienilmastosta Luoteis-Kymenlaaksossa ja siihen rajoittuvilla alueilla. MSc thesis, University of Helsinki

Taarna J (2000) Kouvolan kasvisto. Kouvolan kaupungin ympäristösuojelutoimiston julkaisuja 3/2000:1–174, maps

Taarna J (2007) Kasvien lisälöytöjä ja tuhoutumisia Kouvolassa 2000-luvun alkuvuosina. Kouvolan seudun kansanterveystyön kuntayhtymä, Kouvola

Teitti J (1555–1556) Jakob Teitts klagomålsregister emot adeln i Finland. Microfilms 215j and 215k. The National Archives of Finland, Helsinki. (Manuscript)

Tikkanen M, Oksanen J (2002) Late Weichselian and Holocene shore displacement history of the Baltic Sea in Finland. Fennia 180:9–20. CD. https://fennia.journal.fi/article/download/3760/3551

Chapter 3
Results

Abstract The ages of the luminescence dating of six sand samples from the territory of the current Kouvola town are presented. The oldest dated age, about 63 ka as the minimum, describes the sedimentation of sand on the bottom of a glacial stream appearing now as the glaciofluvial Välikangas sand esker east of the Kymi River. The ages of the samples from Kaurisselkä, Pakinkangas and Raavuori sand ranging from about 12.1 to 9.0 ka are estimates of the depositional time. The Koria sand plain was deposited with water which evidently burst through the outer Salpauselkä ridge 7.61 ± 0.46 ka ago according to luminescence dating. In the Koria sand, strong streaming occurred between 3.6 and 2.0 cal ka BP and downstream from there, the water flooded frequently for which radiocarbon ages were presented. Raised beaches of the X sea after the Ancylus Lake were formed since 7.4 cal ka BP based on radiocarbon dating of three wood charcoal samples.

Keywords Allium schoenoprasum · Biofilm on rock · Floods · Elimäki · Holocene · Ice Age · Jasione montana · Kaurisselkä · Koria · Kouvola · Kymi River · Lotus corniculatus · Pakinkangas · Pärnäjoki River · Phragmites australis · Pleistocene, · Raavuori · Välikangas · Veronica spicata

3.1 Luminescence Dating

The results of luminescence dating for the six samples are presented in Table 3.1. Palaeodose (D_e) distributions for all samples are provided in Chap. 6. Samples NCL-7320041, NCL-7320042, and NCL-7320043 yielded very late Pleistocene to early Holocene ages, ranging from *c.* 12.1 to 9.0 ka. These ages are judged to be fairly reliable estimates of the depositional time, although we caution that the low dose recovery ratio suggests the ages may be underestimated by up to *c.* 10%. For samples NCL-7320044 and NCL-7320045, 100% of grains were above the

H. Ahokas et al., *Paleoshorelines and Their Relict Littoral Plants*, SpringerBriefs in Earth Sciences, https://doi.org/10.1007/978-3-032-17723-0_3

grain-specific saturation threshold (or 2D0 value), so these are minimum ages only of 56 and 63 ka. Overall, the dose rate values of the samples are relatively high, and high dose rates shorten the applicable age range of luminescence dating because they cause the grains to reach saturation more quickly. Sample Shfd161169 returned the youngest age of *c.* 7.6 ka.

3.2 Occurrences of Relict Littoral Plant Species in the Study Region

Of the four plant species, sheep's-bit, spiked speedwell and trefoil grow in warm and relatively dry habitats, usually well exposed to the sunshine (Figs. 3.1, 3.2, 3.3 and 3.4). The chive grows rarely on inland rocks in thin soil, but has its special association with a temporarily drying biofilm of nitrogen-fixing cyanobacteria (Fig. 3.1; Ahokas 2018). For common spreading history and some common environmental requirements, usually two or three of these species grow in the vicinity of each other, confirming the status of an ancient inland shore. The chive is more resistant than the others to saline and, thanks to its bulbs, to drought, but also to excess water. All four species have already occurred on the ancient shores of the Yoldia Sea (Fig. 1.1).

Though here the observations about reed are given only in the Saarela farm (Fig. 3.5a, b), this original species has fairly common occurrences on arable land and forests in patches as remains of the past seashore vegetation of the region. Farming does not permit reed to start growing in the field: to establish itself, the reed needs a shore biotype, usually starting from rhizomes brought by water, but rarely from seeds in well-lighted littoral sites (Luther 1950, 1951).

3.3 Ahkoja Area, Floods Before and After the Appearance of the Kymi River

The Ahkoja-L-1 (Shfd16169) luminescence sample was well bleached before burial and suggests the sand to have moved at 7.61 ± 0.46 ka ago to the site with water from the Ropakonmäki sand esker, a distance of 1.49–1.85 km.

As shown in Table 3.2, the radiocarbon ages of the samples Koria_sandur (Beta-688480) and Ahkoja-3 (Hv-24190) range from 3.568 to 3.177 cal ka BP. At Saarela-6 (Hv-24402), the sample dated to 3.216–2.784 cal ka BP. At Kengäsaho-1 (Su-2985), an age of 2.357–1.888 cal ka BP was returned, whilst at Ahkoja-5 (Beta-469432) the sample dated to 2.306–2.001 cal ka BP. At Simolanaho-1 (Su-3077), the sample dated to 2.321–1.940 cal ka BP. These ages could have resulted from major floods of the proto-Kymi or Kymi through the Ahkoja area. The samples Koria_sandur (Fig. 3.6a, b), Ahkoja-3 and Ahkoja-5 were taken from fairly

Table 3.1 Luminescence dating of sand grains from the bottom of the Koria sand plains (Ahkoja-L-1) on clay plain, from three re-deposited sand or gravel samples (Kaurisselkä-L, Pakinkangas-L, and Raavuori-L), and two samples (Välikangas-1-L and -2-L) found evidently to be deposited in a glaciofluvial esker during the mid-Weichselian Ice Age in the Pleistocene. All sampling sites are in the current inland. The overdispersion value shown in parentheses is that once a single outlier aliquot is removed

Sample name	Laboratory number	Latitude longitude	Altitude (m)	Sampling depth (m)	Palaeodose, D_e (Gy)	Overdispersion, OD (%)	Dose rate (Gy/ka)	Age (ka) Mean ± SE
Ahkoja-L-1	Shfd16169	60° 51′ 06.43″ 26° 35′ 51.05″	55	0.50–0.55	24.1 ± 0.3	10 ± (7)	3.20 ± 0.19	7.61 ± 0.46
Kaurisselkä-L	NCL-7320041	60° 44′ 36.99″ 26° 33′ 08.69″	41	0.85–0.90	61.5 ± 1.6	31 ± 2	5.33 ± 0.21	11.5 ± 0.4
Pakinkangas-L	NCL-7320042	60° 41′ 38.39″ 26° 31′ 54.83″	36	0.65–0.70	44.4 ± 1.1	43 ± 3	4.69 ± 0.20	9.5 ± 0.4
Raavuori-L	NCL-7320043	60° 46′ 31.04″ 26° 35′ 16.06″	56	0.40–0.45	64.2 ± 1.6	35 ± 4	6.74 ± 0.24	9.5 ± 0.4
Välikangas-1-L	NCL-7320044	60° 39′ 46.97″ 26° 53′ 19.45″	35	1.58–1.63	317 ± 9.2	44 ± 6	5.63 ± 0.21	≥56 ± 2.7
Välikangas-2-L	NCL-7320045	60° 39′ 46.97″ 26° 53′ 19.45″	35	1.33–1.38	337 ± 10	45 ± 5	5.31 ± 0.20	≥63 ± 3.0

Fig. 3.1 Small meadow-patches (**A**) tightly grow the highly nitrogen-bound species, the relict chive (*Allium schoenoprasum*). Chive has an association with the temporarily moist, black stream, maintaining nitrogen-fixing cyanobacteria on the rock. Level between 35 and 43 m asl on an ancient shore of the X sea stage. The intact lichens show that humans have not worn the rock. Sipulikallio, Ruotsinpyhtää (Loviisa). June 20, 2010

Fig. 3.2 Relict spiked speedwell (*Veronica spicata*) on thin soil on rock with a warm, intact habitat. The site also has trefoils and sheep's-bits. 97 m asl. Ancient shore of the Yoldia Sea not worn by humans. Haukkavuori, Artjärvi (Orimattila). July 10, 2005

Fig. 3.3 Relict sheep's-bit (*Jasione montana*). The site also has chives and trefoils. 63 m asl. Ancient shore of the Ancylus Lake. Laukkakallio, Elimäki (Kouvola). August 3, 2019

Fig. 3.4 Relict trefoils (*Lotus corniculatus*) about 10 m apart from the sheep's-bit of the previous Fig. 3.3, and also 63 m asl. May 27, 2018

deep layers of the western Koria sand plain, which suggests that water had streamed strongly 3.65–2.00 cal ka BP. Figure 3.7 shows spherical grains of clay and sand from the depth of 50–55 cm, rounded by the streaming reaching the bottom of the sands.

The small tributaries of the Ahkoja and Pyteränoja join the Kymi river at 139° and 152° angles, respectively, i.e., against the flow of the latter (Fig. 1.2). This shows that the Ahkoja and Pyteränoja started originally as branches. Ahkoja and Pyteränoja have significantly more branching gorges on the proximal than the distal sides (Ahokas 2003), which coherently proves streaming of the floods from N or NW to these present-day tributaries.

Fig. 3.5 (**a**) Relict reed (*Phragmites australis*) as the perennial weed emerged from its rhizomes from season to season in the field of the Saarela farm, which had grown different crops at least since 1792. 40 m asl, 60° 45.134′ N, 26° 31.292′ E, before the heading stage, on June 28, 1997. (**b**) The same site as in (**a**) with heading reed on August 4, 2021, after the afforestation, which can be seen with the open-access aerial photographs (NSL aerial photographs 2023). The particular survey for the reed in the Saarela farm in 1994 showed two, about 0.86 and 1.51 ha survival stands, which were confined to 39 á 41 m asl. This raised the hypothesis of a past existence of the shore at this level after the Ice Age. The bolder lines and rounded, washed gravel from angular moraine particles occurring at *c.* 40 m asl also suggested long-lasting littoral effects in Saarela and the neighborhood. The Kaurisselkä and Pakinkangas samples for dating were taken, respectively, 1.88 km ESE and 6.44 km S of this Saarela site. The Kaurisselkä sample is from the same level. Water of this ignored sea also covered with about 10 m the site of the Ankkapurha rapids, less than 17 km towards the ESE. These locations are marked in Fig. 1.1

The widely occurring charcoal-rich topsoil in the Koria sand plains (Fig. 3.6a) had been radiocarbon dated earlier. The area has been under the *kytö* burning cultivation, leaving the charcoal in the topsoil. The map from the year 1777 (Harju 2012) shows the area to have been grassland, which was a typical succession in the *kytö*-burning management, while the *kytö* appellative also occurs in more particular old maps here. The *kytö* was a common tillage practice in many regions in Finland in the eighteenth century (Ahokas 2012). The mapping in 1777 permits the multiple peaks of calibration of Ahkoja-1 (Su-3337, Table 3.3) radiocarbon age to be deciphered to 1666–1784 years BP ($p = 41.8\%$). The use of such other evidence was suggested for calibrating the last few centuries from the occasionally numerous alternatives (Pauly 2020).

3.4 The Raavuori Rock with a Major Raised Beach in Inland

The luminescence sample of Raavuori (NCL-7320043) was taken from the upright wall of a small sand pit and from the 56 m level which corresponds the central bottom of the 3–4 m high, 235 m long, and 25–35 m broad raised beach (Fig. 3.8a). The sampled colluvial gravel was 0.48 km ENE of the center of the raised beach (Fig. 3.8a, b). The luminescence age 9.5 ± 0.4 ka (NCL-7320043) of the Raavuori-L sample dates a long-lasting sea high stage in these positions, which is younger than

Table 3.2 Radiocarbon (^{14}C) dating of the frequently occurring floods at and downstream from the Koria sand plains

Sample type	Distance (km) from the Koria sands to direction (°)	Altitude (m)	Sampling depth (cm)	Sample name Sampling year *Previously published (Ahokas 2003)	Laboratory number	^{14}C age a BP	Calibrated age, cal a BP	U and *Th* (ppm) in the area enabling the ^{14}N (n,γ) ^{14}C reaction
Rounded sand aggregates and clay particles (Fig. 3.7)	0	55	50–55	Koria_sandur 2016	Beta-688480	3370 ± 40	3649–3491	4.41–5.03 *10.2–11.8*
Sand–clay interface under sand	0	56	46–50	Ahkoja-3 2000*	Hv-24190	3170 ± 80	3568–3177	4.41–5.03 *10.2–11.8*
Charcoal in sand on clay	0	55	75–85	Ahkoja-5 2016	Beta-469432	2150 ± 40	2306–2001	[a]
Charcoal in fluvial sand on river sides	6.2 214°	51	15–25	Pärnäjoki-1 2000*	Hel-4438	410 ± 80	553–294	[a]
Gyttja clay under peat	11.3 202°	40	26–30	Suppamäki-1 1998*	Su-3078	1530 ± 60	1529–1309	3.84–4.4 *15.6–17.6*
Gyttja clay under peat	11.4 206°	43	18–21	Simolanaho-1 1998*	Su-3077	2130 ± 70	2321–1940	2.8–3.31 *10.2–11.8*
Sand–clay interface under sand	11.5 207°	52.8	60–61.5	Saarela-6 2001*	Hv-24402	2870 ± 80	3216–2784	2.8–3.31 *13.7–15.5*
Gyttja clay under peat	11.9 193°	38	23–26	Kengäsaho-1 1998*	Su-2985	2170 ± 100	2357–1888	4.41–5.03 *10.2–11.8*
Gyttja clay under peat	13.6 168°	39.5	*c.* 30	Kaukasuo-1 1998*	Su-3079	980 ± 80	1058–727	2.23–2.79 *4.89–8.17*

Samples indicated with a * were previously published in the cited reference

[a]Charcoal samples ignored

Fig. 3.6 The western part of the Koria sand plains excavated through the sand and a few decimeters into the clay underneath for setting a new heating pipe. (**a**) The black upper layer on the excavated surface (marked with yellow rings) is charred remains from the *kytö* fire cultivation era in the eighteenth century. Sept. 25, 2016. (**b**) Tube with black polypropylene foam stopper for Ahkoja-L-1 (Shfd16169) luminescence sample at the bottom of the sand (Table 3.1). The sample for Beta-688480 radiocarbon dating was from the same depth as for the luminescence, while that for Beta-469432 was from *c.* 10 cm deeper (Table 3.2). The dark, moist clay is visible under the sand. Oct. 2, 2016

Fig. 3.7 The Koria sand plains. Rounded clay particles and aggregates of sand from 1 to 6 mm in diameter were most likely formed by streaming water. Scale in millimeters. Sampled from the luminescence dating depth (Fig. 3.6b). The rounded grains were selected, cleaned from loose sand, and reduced with a steel device (Ahokas 2023) to powder, which was pretreated to eliminate carbonate and processed for the AMS dating. Age 3.649–3.491 cal ka BP (Beta-688480) in Table 3.2

the generally assumed Ancylus transgression. This sea high stand continued after the obtained 9.5 ka age, as evidenced by the overlying deposits that accumulated the raised beach in the head of a bay.

Table 3.3 Grain size distributions as mass percentage from depths in centimeters below the surface of the mineral soil in Koria sands, Raavuori, and Pakinkangas B site. ^{14}C = radiocarbon sample. Lumin. = luminescence sample

	Koria sand plains		Raavuori			Pakinkangas B		
Grain size (Ø mm)	^{14}C site of Ahkoja-1 14–20 cm	Lumin. site 46–50 cm	Lumin. site 25–30 cm	Lumin. site and depth 40–45 cm	Lumin. site 45–50 cm	Lumin. site 20–25 cm	Lumin. site and depth 65–70 cm	^{14}C site 38–42 cm
>6.3	0.0[a]	0.0	19.4[b]	0.6	3.8	3.6	4.1	0.2
≤6.3 >2.8	0.3[a]	0.1	59.7[b]	18.4	14.1	8.9	0.8	3.6
≤2.8 >2.0	0.6[a]	0.4	2.6[b]	9.1	20.5	5.6	0.5	2.3
≤2.0 >0.6	5.8[a]	6.7	3.5[b]	25.0	52.2	27.9	2.5	15.7
≤0.6 >0.063	86.0[a]	89.4	14.7[b]	43.3	9.4	52.9	91.5	77.6
≤0.063 >0.02	6.8[a]	3.4	0.1[b]	0.09	0.06	1.0	0.5	0.5
≤0.02	0.5[a]	0.2	0.06[b]	0.0	0.0	0.0	0.05	0.0

[a] The fractions contained charcoal apparently from the past *kytö* burning cultivation; Ahkoja-1 (Su-3337, 150 ± 50 BP) calibrated to the seventeenth or eighteenth century (see text)

[b] Rusty appearance from podsolization likely caused by humus acids

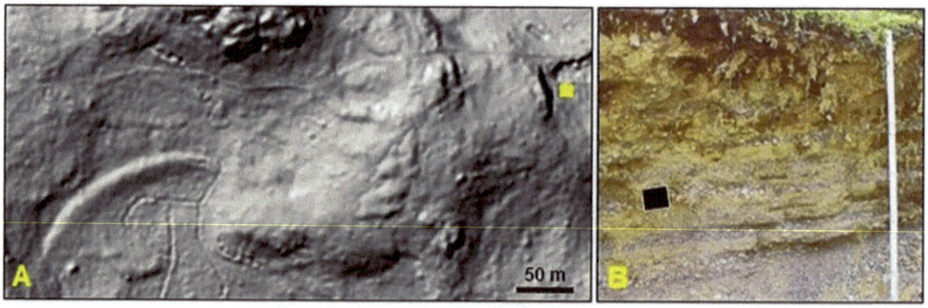

Fig. 3.8 The Raavuori rock. (**a**) The 235 m long arch of the 3–4 m high raised beach on the left side, and the luminescence (NCL-7320043) sampling site (yellow dot on the right edge) on the bottom level of the raised beach. View from MapSite (2022–2024). (**b**) Taking the luminescence sample Raavuori-L (the black square) from the level corresponding to the bottom of the central raised beach. Centimetric strata of fine and coarse sand alternate. The sampling depth is 40–45 cm. Sept. 12, 2019

The sampling level of Raavuori is also close to that of the Ahkoja luminescence sampling site on the Koria sand plains, 55 m asl, 8.7 km N of the Raavuori site. The luminescence age of the Raavuori-L sample (9.5 ± 0.4 ka) is quite distinct from the age of the Ahkoja-L-1 luminescence age (7.61 ± 0.46 ka; Shfd16169; Table 3.1). The Ahkoja-L-1 sample originated about 1 m lower in topography than the Raavuori-L sample. The Koria sand plains with discharge spreading water and moving fine sand ≥1.49 km is different from that of the Raavuori sampling site, which contained sharply stratified colluvial gravel (Table 3.3). This is also interpreted to indicate that strong, meter-high waves were driving large clastic sediment to the raised beach in the past ancient bay on the Raavuori (Fig. 3.8a).

3.5 Kaurisselkä and Pakinkangas Eskers with Inland Shorelines of the Past Seas

The sand of Kaurisselkä and Pakinkangas was spread by the Yoldia or another earlier sea laterally from their approximately N–S directed glaciofluvial sand eskers. This happened 11.5 ± 0.4 and 9.5 ± 0.4 ka ago (NCL-7320041 and NCL-7320042, Table 3.1). Under the dated levels, there occur one or more clay strata of ±1 cm thick (Figs. 3.9c and 3.10b). This is taken to indicate that, the westwards and eastwards spread of sand was later under multiple shore formations which left on the eastern slope of the Kaurisselkä esker, and on the southeastern slope of the Pakinkangas esker, seven ancient raised beach levels. Radiocarbon dates from wood charcoal from the proximal side of three of them (Figs. 3.9a and 3.10a) date some of these shoreline stages (Table 3.5). The Kaurisselkä radiocarbon sampling, just over 40 m asl, also dates the ancient shore 1.88 km WNW in Saarela with the old reed stand (Fig. 3.5a, b, Table 3.5). On the eastern Kaurisselkä, the seven distinguishable raised

Fig. 3.9 (**a–c**) The Kaurisselkä site is about 0.5 km west of the Kaurisselkä esker with raised ancient beaches. (**a**) The upper (U), middle (M), and lower (L) beaches with the luminescence (NCL-7320041) sampling site (yellow dot) and the wood charcoal site with a black dot (Beta-532636). Hillshade and the 40 m asl isopleth from MapSite (2022–2024). (**b**) View from NE towards the parallel upper (U) and middle (M) raised beaches, the latter being on the same level as the Saarela site 1.88 km WNW, with the ancient occurrence of reed (Fig. 3.5a, b). The distance between the tops of U and M is 31 m. The heights of these raised beaches are ±2 m. May 16, 2025. (**c**) Taking the luminescence sample Kaurisselkä-L from the bottom of the NE side of the M beach sand at 0.85–0.90 m under the surface, above clay, with a 5 × 5 cm tube with black polypropylene foam stop. Sept. 3, 2019

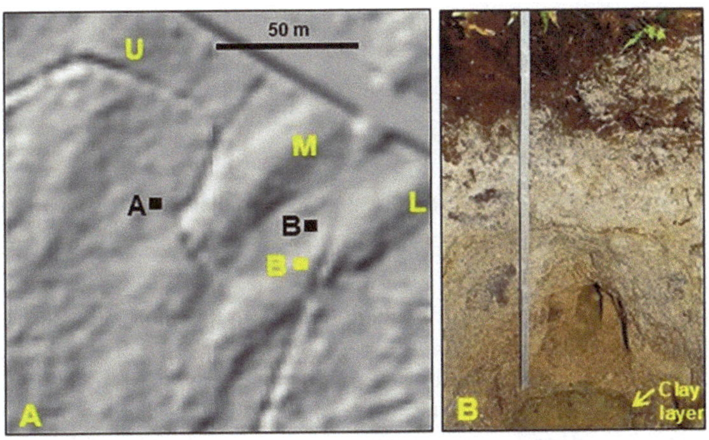

Fig. 3.10 (**a, b**) The Pakinkangas raised beaches on the eastern side of the esker. (**a**) The luminescence (NCL-7320042) sampling site (yellow dot B), and the two wood charcoal sites, black A = Beta-532637, and black B = Beta-532638, in relation to the upper (U), middle (M), and lower (L) raised beaches from MapSite (2022–2024). (**b**) Taking the luminescence sample Pakinkangas-L from the bottom of the beach sand at 0.65–0.70 m under the surface, above clay layers. Sept. 12, 2019

beaches are above the 37.5 m isopleth (MapSite 2022–2024). Among the multiple shore levels both in Kaurisselkä and Pakinkangas, there must also be remains of transgressions, which is suggested by the about 3360 years younger age for the upper Pakinkangas-A sample compared with the age for the lower Pakinkangas-B (Table 3.5).

3.6 Raised Beaches on the Prior Glaciofluvial Sand, the Present Välikangas Esker

The Välikangas esker is *c.* 3 km long, roughly in N–S direction, and typically *c.* 0.6 km crosswise (Häyry et al. 1961), which direction and morphology suggest it originated from a glaciofluvial river. There, the water ran southwards during the melting of the continental ice, and the stream also left fragments of eskers both north and south of the Välikangas sampling site. The south or southeastern slope of the Välikangas esker hosts seven ancient raised beach levels between 32 and 37 m asl (Fig. 3.11a).

Clearly, disturbance of the sands of different ages had occurred, the older being above the younger (Table 3.1). The undisturbed sand layers (Fig. 3.11b, c) in the sampling depths show that this disturbance had happened before the stratification, and is explainable by the southern slope on which the streaming water in the glaciofluvial river caused the pedoturbation downslope (Fig. 3.12). The sand grain size distributions from the 5 to 163 cm depths are fairly uniform, only the layer 5–15 cm having a slightly coarser distribution (Table 3.4).

3.7 Radiocarbon Dating of the Ancient Shores of About 40 m asl from Wood Charcoal

Dated wood charcoal from the upslope side of the raised beaches in Kaurisselkä and Pakinkangas (Figs. 3.9a and 3.10a) is presented in Table 3.5. These samples were aimed at dating the existence of the shore with the waves washing incidental charcoal over a beach berm, burying it on the bottom of the tiny valley there, while the luminescence ages rather date the formation of the sandy bottom of the raised beach. The distance between Kaurisselkä and Pakinkangas sites is 5.7 km, and their measured annual land upheaval difference is *c.* 0.15 mm (Fig. 1.1). The difference in 7400 years would predict a 1.11 m upheaval surplus to Kaurisselkä, and also to the Saarela site (Figs. 1.1 and 3.5a, b), compared with Pakinkangas. The radiocarbon ages of Kaurisselkä, 6.88–6.68 cal ka BP, and Pakinkangas-B, 7.42–7.19 cal ka BP, could mean that Pakinkangas-B was a seashore somewhat earlier. The Pakinkangas-A radiocarbon age 4.06–3.83 cal ka BP from a level about 0.5 m higher than

Fig. 3.11 (**a–c**) The Välikangas site on the border of the sand-pit. (**a**) The esker has seven distinct raised beach levels on the slope exposed towards S-SE, shown with the numbers 1–7. The yellow arrow shows the sampling site on the highest raised beach presented in (**b**, **c**). Hillshade from MapSite (2022–2024). (**b**) View from east showing the points of NCL-7320044 and NCL-7320045 samplings below and above the three clay-rich layers. The arrows show 135 and 161 cm levels below the top. June 22, 2019. (**c**) The squared sample tubes for luminescence are fitted into the sand with black polypropene foam stoppers. The clay layers are between

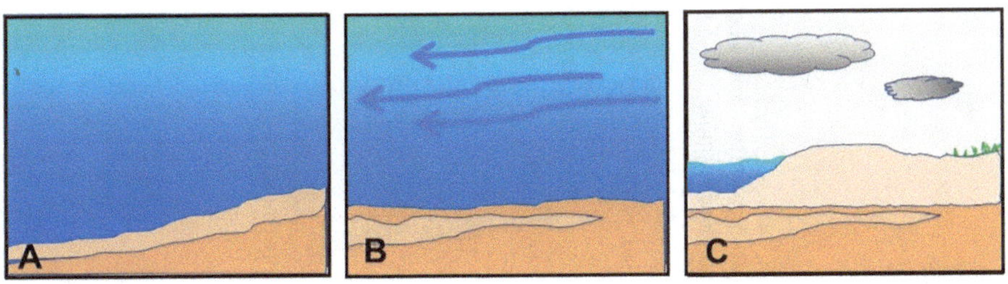

Fig. 3.12 (**a–c**) Semantic illustration of the possible series of events leading to the pedoturbation or old-up stratification of sand on the bottom of the inclined glaciofluvial river on the present southern slope of the Välikangas esker (Fig. 3.11b). The darker soil depicts the older (≥63 ka), and the lighter, the younger (≥56 ka) sand. The upper part of the raised beach is a mixture of redeposited age classes (in **c**)

Pakinkangas-B could be a later, transgression shore (Table 3.5). The eastern slope of the Kaurisselkä esker, *c.* 0.5–0.6 km eastwards of the sampling site, has evidence for seven raised beaches, among which there might be exact equivalents to those seven beaches on the SE exposure in the Pakinkangas esker. Additional radiocarbon

Table 3.4 Grain size distributions as mass percentage from depths in centimeters below the surface of the mineral soil in Kaurisselkä and Välikangas. ^{14}C = radiocarbon sample. Lumin. = luminescence sample

	Kaurisselkä				Välikangas			
Grain size (Ø mm)	Lumin. site 15–20 cm	Lumin. site and depth 85–90 cm	^{14}C site 70–75 cm	5 m S of top of the raised beach 20–25 cm	Top of the raised beach 110–120 cm	Lumin. site 5–15 cm	Lumin. site, upper depth 133–138 cm	Lumin. site, lower depth 158–163 cm
>6.3	0.0	0.0	0.0	0.2[a]	0.0	0.7	0.0	0.0
≤6.3 >2.8	0.3	1.1	0.3	0.8[a]	0.4	0.8	0.0	0.4
≤2.8 >2.0	0.7	0.8	0.6	1.3[a]	1.2	0.7	0.0	0.5
≤2.0 >0.6	21.3	18.0	15.3	24.6[a]	9.6	9.1	0.4	8.0
≤0.6 >0.063	76.9	79.2	83.0	71.7[a]	56.9	87.5	89.0	76.2
≤0.063 >0.02	0.8	0.7	0.7	1.2[a]	28.7	1.1	10.1	13.5
≤0.02	0.05	0.05	0.1	0.2[a]	3.1	0.1	0.5	1.3

[a] Rusty appearance from podsolization likely caused by humic acids

Table 3.5 Radiocarbon (^{14}C) dating of the ancient raised beech shores in Kaurisselkä and Pakinkangas eskers with charcoal samples

Sample type	Altitude (m)	Sampling depth (cm)	Sample name Sampling year	Laboratory number	^{14}C age a BP	Calibrated age cal a BP
Charcoal in sand under peat and sand on clay	41	58–63	Kaurisselkä 2019	Beta-532636	5950 ± 30	6881–6676
Charcoal in muddy sand on clay	37	22–28	Pakinkangas-A 2019	Beta-532637	3600 ± 30	4061–3832
Charcoal in sand under peat and sand on clay	36.5	33–42	Pakinkangas-B 2019	Beta-532638	6390 ± 30	7423–7186

dating would be needed to validate this. The water of these sea stages also covered the Ankkapurha rapids area (Fig. 1.1) at a distance of about 15 km ESE and E of Kaurisselkä and Pakinkangas, respectively, so evidently still 4.06–3.83 cal ka BP.

References

Ahokas H (2003) Major discharge and floods of the Kymi river (Kymijoki) over the Salpausselkä I esker to coastal South Finland traced by ^{14}C datings, unequal land upheaval, oral tradition and onomastics. Kave, Helsinki

Ahokas H (2012) Crop evolution under fire. Interdiscip Biol Agric Linguist Antiq 5:1–194

Ahokas H (2018) Ruoholaukan uusi löytöpaikka muinaisrannalla Kymenlaaksossa ja mustan biofilmin merkitys. Summary: Allium schoenoprasum on the prehistoric lake shore at a recently discovered inland site in SE Finland. Lutukka 34:104–111. https://tietopankki.luomus.fi/lehdet-ja-kirjat/lutukka/

Ahokas H (2023) Thirty-three previously unknown meteoritic craters of diameter from 5 to 100 m in western Kouvola, Finland from a swarm of impactors in the Holocene. Kave, Helsinki. https://www.impact-structures.com/finland

Harju E-S (2012) Kuninkaan kartasto Etelä-Suomesta 1776–1805. Suomal Kirjall Seur Toimit 1373:1–477

Häyry P, Toivonen T, Vuorinen J, Sillanpää M, Mansala K, Antikainen J (1961) Juurikorpi 3024 11. Maataloudellinen maaperäkartta 1:20 000. Maatalouden tutkimuskeskus and Maanmittaushallitus, Helsinki. (Soil map)

Luther H (1950) Beobachtungen über die frutifikative Vermehrung von Phragmites communis Trin. Acta Bot Fenn 46:1–18. Acta Botanica Fennica 1950–1951:46–48: Free Download, Borrow, and Streaming: Internet Archive

Luther H (1951) Verbreitung und Ökologie der höheren Wasserpflanzen im Brackwasser der Ekenäs-Gegend in Südfinnland. Acta Bot Fenn 49:1–231, appendices

MapSite (2022–2024) The National Land Survey of Finland, Helsinki. maanmittauslaitos.fi/en/e-services/mapsite

NSL aerial photographs (2023) [Public aerial photographs] kartta.paikkatietoikkuna.fi

Pauly M (2020) Challenges in accuracy over the last few decades. Radiocarbon dating of the Modern Era. Beta Analytic Inc., Miami, FL. (E-mail)

Chapter 4
Discussion

Abstract Glaciofluvial sand sedimentation probably started millennia years earlier than 63 ka dated from the Välikangas sand. The melting in the Pleistocene end resulted in the "ice-free" period during the Ice Age in parts of Fennoscandia, which was a requirement for the occurrence of mammoths. Post-glacial shores were occupied with littoral plants which, after land upheaval, sometimes survived over millennia years in inland habitats where their special ecological requirements happened to be fulfilled. Relict plant populations were saved on shores in the region from Yoldia Sea (with the sea level between 78 and 100 m asl), Ancylus Lake (between 50 and 60 m asl) and the undetermined X sea stage (between 35 and 43 m asl) which prevailed before the Littorina Sea (below 28 m asl). Where the altitude on a sandy slope is high enough, seven raised beaches at the most were detected on the ancient X sea. An age estimate of the Tyynelä rock art on the X sea shore level in Elimäki was obtained from a radiocarbon dating. The floods and ancient channels of the developing Kymi River were remembered in the oral tradition. Ice blockage in the rivers could rapidly be generated by supercooled water. Observations of supercooling in the Kymi River appear in literature. Toponyms on firm land with 'island' as an appellative are memories of past floods. Dwelling on elevated places was remembered. The ages of Kymi and Vuoksi Rivers, both of which burst out through the Salpausselkä ridge, are compared introducing some previously ignored radiocarbon dates.

Keywords Climate changes · Floods · Ice blockages · Ilmajärvi · Kymi river · Ladoga transgression · Land upheaval · Middle Weichselian melting · Prehistoric islands · Prehistoric shipwrecks · Rock arts · Sea level changes · Supercooled river water · Vuoksi river

H. Ahokas et al., *Paleoshorelines and Their Relict Littoral Plants*, SpringerBriefs in Earth Sciences, https://doi.org/10.1007/978-3-032-17723-0_4

4.1 The Ice-Free Middle Weichselian

There are plenty of publications on the maximal extent of the Weichselian Ice Age, Last Glacial Maximum between *c.* 20 and 12 ka ago (Hughes et al. 2016; Anon. 2024–2025). The minimum ages obtained in this study for Välikangas sand, 63 and 56 ka (Table 3.1), would match a model that the older Weichselian ice was then melting in the inner SE Finland-to-be and caused large glaciofluvial rivers. Glaciofluvial sand sedimentation may have occurred even much earlier than 63 ka, also in Välikangas, since it was dated in Finnmark, northeastern Norway, a glaciofluvial sand age of 83.5 ± 6.6 ka (Goodship et al. 2025). Starting from scattered observations, the Middle Weichselian era has lately been shown to have been "ice-free" in parts of Fennoscandia, e.g., based on dating of remains of mammoths, the indicator of the extincted mega-fauna: in the Finland-to-be from *c.* 32 to 16 uncal ka BP radiocarbon ages (Ukkonen et al. 1999: Table 1), i.e., 37–19 cal ka BP (Ukkonen and Mannermaa 2017: p. 209). Most of the mammoth remains from Sweden were dated to 38–24 ka ages (Ukkonen et al. 2007).

The existing downslope has directed the appearance of glacial rivers towards the future coast. The N-S elongated Välikangas esker (Fig. 3.11a) was formed during the Middle Weichselian as glaciofluvial bottom sand. The dated sand samples give an age estimate to the glaciofluvial streaming on Välikangas, ≥63 and ≥56 ka (NCL-7320045, NCL-7320044; Table 3.1). Streaming force on Välikangas has apparently caused pedoturbation, or old-up re-layering of the sand, because of the sand slope on the site, when the enforced streaming of water above had pushed sand downwards, as semantically illustrated in Fig. 3.12. A bioturbation by flora or fauna (Bateman et al. 2007) did not cause the old-up turn of the sand at the depth between 1.33 and 1.63 m in Välikangas, where the sampled strata also looked intact.

4.2 Post-glacial Shores Were Populated with Littoral Plants Which Sometimes Survived in Inland Habitats

Chive, sheep's-bit, spiked speedwell and trefoil are hemerophobic species and have survived in anthropologically undisturbed areas of Finland: see e.g. the undisturbed lichen-covered rocks in Figs. 3.1 and 3.2. The species have survived as indicators of their original habitats. Trefoil additionally appears as an apophyte on roadside gravel, where application of deicing salt since the 1960s has formed a strip resembling seashore ecology. Further, competitive trefoil obligatorily requires its *Rhizobium* species in the soil and must have been seeded by soil or seeds carrying the symbiont bacteria.

The granites of southeastern Finland are, when moist, vulnerable to maintaining cyanobacterial biofilms, which have been encountered in St. Petersburg, Russia, with the constructed rocks imported from SE Finland (Panova et al. 2014). The reason for the survival of chive on inland rocks seems to depend on the supply of nitrogen from the cyanobacterial biofilm on sunny rock surfaces where seeping moisture is received (Fig. 3.1; Ahokas 2018). Chive in its original seashore habitats is nitrophilous (Eklund 1958), thrives in the splash zone, and benefits from a high supply of nitrogen fertilizers in cultivation (Hartmann 1966; Suojala 2003). The seed germination of the endemic chive originating from above timberline in the Iberian Peninsula responds to cold stratification and warm germination temperature (Giménez-Benavides et al. 2005), whose requirements may lead to a seed bank in the soil. The perennial chive also survives with its bulbs during dry seasons.

In Eklund's definitions (1958), the perennial trefoil is suprasaline and supralittoral on seashores. Because it produces from 30% to 56% of hard seeds (Pohjakallio 1941), trefoil tends to form seed banks in the soil. Trefoil grows on fairly dry, weathered rocks and has a deep woody taproot. It is common on the shores of the Gulf of Finland (Kujala and Ulvinen 1964). Its rhizobial symbiont guarantees nitrogen supply, making the trefoil a pioneer species on drained sandy shores. The yield of trefoil increased about three times, and overwintering benefited from the symbiosis in the experiments (Pohjakallio 1941). Efficient rhizobia or trefoil itself evidently needs a level of ≥900 mg/L of soluble Ca and a pH of ≥5 in its natural habitats in Finland (Ahokas 2006).

Trefoil, chive and sheep's-bit flower in this order, and may have mutually maintained pollinating insects in a common habitat in the pioneer shore vegetation. Seed dormancy may have guaranteed the survival of the annual or biennial sheep's-bit, as it seems to be absent from its known habitats in some seasons. Trefoil and chive are heliophilous, while sheep's-bit is moreover a thermophilous species and these species entered the flora of Denmark during the Alleröd interstadial (Iversen 1954). Their occurrence and survival until now in warm and sunny sites on the ancient Yoldia and Ancylus Sea shores in the inland of Finland is in accordance with their prehistoric spread during the warm climate periods after the Weichselian Ice Age. Such warming evidently occurred in the Finland-to-be between 11.7 and 9.5 cal ka BP (see Väliranta et al. 2015). The current climate type seems to permit the sheep's-bit to spread on littoral rocks in the SW coast of Finland (Kunttu 2019).

In the inland of Finland, the northernmost occurrences of sheep's-bit and spiked speedwell are in the North Karelian province (Lampinen and Lahti 2019), which was ice-free already during the Yoldia Sea, *c.* 11.2–10.2 cal ka BP (Tikkanen and Oksanen 2002; Ancient shorelines 2013; Ojala et al. 2013). Spiked speedwell is rare in Finland nowadays, but occurs principally near the sea (Lampinen and Lahti 2019) and especially on rapakivi granite rocks. It needs summer warmth, sunny, sandy habitats, even dunes (Hartl 1966), which make it a pioneer species on uplift shores. It also requires some alkali with or without lime; sea water pH in the Gulf of Finland is typically >8 in the coastal (Ahokas 1992) and central water (Almén et al. 2017). Spiked speedwell is adapted to relatively dry soils, because water holding capacity over 70% of lime soil seems to retard its growth (Misra and Tyler 1999). Spiked speedwell, chive and trefoil currently occur together on three small isles, viz. Ristisaari (60° 18.7′ N, 26° 48.6′ E), Kilpisaari (60° 16.8′ N, 27° 18.1′ E), and Mustamaa (60° 26.1′ N, 27° 32.7′ E) in the Gulf of Finland, south of the SE coast (Lampinen and Lahti 2019).

Eight km beyond the left (western) border of Fig. 1.1, the occurrence of chive on 40 m asl (60° 26.360′ N, 25° 22.838′ E) must be ancient, as the species has named the concerned village as Laukkoski ('Chive-rapids'), in the Pornainen municipality, and *laukka* further named some historical land parcels in Laukkoski. About 100 m eastwards of the current occurrence of chives in Laukkoski, the hillshade map (MapSite 2022–2024) shows remains of four ancient, raised shore levels between 35 and 45 m asl. Archaeologist Pesonen (2003) deduced the 40 m level here to have been inhabited *c.* 9.55 ka BP.

4.3 Sea Shores Between the Recognized Ancylus Lake and the Litorina Sea in the Kymi Dale

Ancylus transgression in the Kymi Dale, in the Hangassuo swamp, occurred, based on radiocarbon ages, between 11.17–10.13 and 10.91–9.54 cal ka BP (Eronen 1976: Hel-661, Hel-664), and 11.88–10.49 cal. ka ($p = 94.7\%$; Hyyppä et al. 1969: Su-38). These ages point to the beginning of the Ancylus era at about 11.5 ka cal BP in the Kymi Dale. The deposition of the "bottom" sands, sampled as Kaurisselkä-L, 11.6 ka (NCL-7320041) and that of Pakinkangas-L, 9.5 ka (NCL-7320042), and the tested shore level of Raavuori-L, 9.5 ka (NCL-7320043) (Table 3.1) fit to the calibrated age limits of the Ancylus from the Hangassuo peat samples. The preserved 235 m long, 3–4 m high, and 25–35 m broad raised beach arch on the Raavuori rock (Fig. 3.8a) has its bottom on the Raavuori-L sampling level and can be regarded as an Ancylus Lake formation at the head of the bay at the time.

The Litorina Sea upper levels were determined by Hyyppä (1937) in part of the region surveyed here. The Litorina maximum shore in Pernaja municipality (*c.* 60° 27′ N, 26° 02′ E) was 28 m asl, and in Pyhtää municipality (*c.* 60° 30′ N, 26° 33′ E) 26–27 m asl (Hyyppä 1937: p. 78). The shipwreck found in Pernaja (Hipping 1817), in the 430-ha Röjsuo swamp (Tynni et al. 1976), was from the Litorina stage, 6.3 cal ka BP or more (Fig. 1.1). Between Pernaja and Pyhtää, the Sipulikallio ('Onion-rock') occurrence of chive (Fig. 3.1) has survived on the shore level of 35–43 m asl, which is between the Ancylus and Litorina levels. In Juurikorpi, *c.* 15 km NE of Pyhtää, the Litorina limit is 27–28 m asl (Hyyppä 1937: p. 96).

The Litorina maximum in Inkeroinen, i.e., on the east shore of the Ankkapurha rapids of the Kymi River (Fig. 1.1), should have been at most 28–29 m asl (Hyyppä 1937: p. 95). Reinholm (1839–) recorded the local people's oral information that a sea had extended to Ankkapurha rapids in the past, while, as shown here, the Ankkapurha area was evidently under water still 4.06–3.83 cal ka BP. In the Reitkalli area, the Litorina maximum is *c.* 27 m asl (Hyyppä 1937: p. 96). There, chives occur 43 m asl on the remote Lautkallio rock (60° 35,608′ N, 27° 03,576′ E) known for decades (Kujala and Ulvinen 1964). The Lautkallio name is an elliptic derivation of Lauk(ka)kallio, 'Chive-rock'. Other changes from Lauk- to Laut-appellatives in Finnish toponyms were given by Lopmeri (1947: p. 98). The rock art shown in Fig. 4.1a represents the human activity on the Litorina shore. The shoreline database

Fig. 4.1 (**a**, **b**) Two examples of the prehistoric rock art south of Ss1, on the SE coast of Finland. (**a**) Carved pictorial writing on the surface of a semi-cave. The nearest logogram is a six-sax, putatively depicting a snake. The measuring band shows 65 cm. About 19 m asl, with the apparent linkage to the nearby prehistoric dwelling site. Kummikirkonkivet, Virolahti, R-K in Fig. 1.1. Found in 2013 by Dr. Hannu Ahokas (2015). Photo: H. Ahokas on Aug. 7, 2013. (**b**) Red-brown rock painting depicting an animal (length 36 cm) looking to the left, and an adorant (height 87 cm, space between the adorant's hands 45 cm), probably painted standing on the ice of the X sea 42–43 m asl (compare with Fig. 4.2). Tyynelä, Elimäki (Kouvola), R-T in Fig. 1.1. Found in 1999 by Mr. Veikko Turkki, M.Sc. Photo: H. Ahokas on Oct. 3, 1999

(Ancient shorelines 2013; Ojala et al. 2013) compiles the observations of the Litorina shoreline.

In the Helsinki region, the highest Ancylus level is now detectable 60 m asl, but the late Ancylus shore levels appear 40–48 m asl in the eastern area of the Helsinki region (Hyyppä 1950: Figs. 33 and 34, pp. 43 and 44). These shores of the Helsinki region are older than those on the corresponding altitudes in the Kymi Dale. This assumption is based on the radiocarbon dated peat bog Metsolansuo in Vantaa *c.* 43.7 m asl (Eronen and Haila 1982), 9.88–9.13 cal ka BP (Hel-1239) and 10.17–9.14 cal ka BP (Hel-1240), and 24 km NE of Hyyppä's observation point (1950), in the Laukkoski dwelling site *c.* 40 m asl, *c.* 9.55 ka BP (Pesonen 2003).

From a previous study (Ahokas 2017: Table 9 and Fig. 83), the radiocarbon dating of wood charcoal (Kraateri E2, Poz-63,298) from a layer in clay soil suggests dry land to have occurred at 9.74–9.53 cal ka BP on the modern 47.5 m asl (0.51 km NNE of the Saarela site; Figs. 1.1 and 3.5). The prehistoric painting on rock in

Elimäki, Kouvola, at the drained Elimäenjärvi lake (Fig. 4.1b; Miettinen 2000; Ahokas 2003), is 5.68 km SW of the Kraateri E2 (Poz-63298) sampling site and 42–43 m asl. Its painting may have occurred at about 40 m sea level, by standing on the ice, because altitudes of rock arts have often been found to relate to ancient sea, lake, and river water levels (Kivikäs 1995; Miettinen 2000). The upright rock surface was probably in contact with the sea level at that time. If so, the age of the painting is between 6900 and 6700 years based on the Kaurisselkä radiocarbon age sampled from the paleoshoreline 6.26 km ENE of the painting (Table 3.5).

4.4 Putative Still-Water Stages Under Ice Cover

The clayey or very fine sand strata at the level 40 m in Kaurisselkä, 36 m in Pakinkangas, and 35 m in Välikangas (Figs. 3.9c, 3.10b, 3.11b, c) suggest that still-water stages existed. They probably lasted over a summer or most of it, or over several years and might occurred during the Younger Dryas, which set in *c.* 12.86 cal ka BP (Punkari and Boulton 1995), or 12.80 cal ka BP (Moore et al. 2023), and ended in the research region, 11.653 cal ka BP (Donner 2010) or *c.* 11.64 cal ka BP (Punkari and Boulton 1995). A freezing period during the Younger Dryas, causing still water under ice, would be consistent with luminescence dating of the samples from above the clay layers in Kaurisselkä and Pakinkangas (Table 3.1). In contrast, older sand with clay was re-deposited on the Välikangas site before and after the putative still-water period.

4.5 Storms, Prehistoric Shipwrecks and Art on Rock

The ancient raised shore formation on the Raavuori and the coarse gravel stratified in centimetres-thick annual layers (Fig. 3.8b, Table 3.3) on its eastern side show that a stormy weather type has prevailed in the past, about 9500 years ago, when the sea level was some below the present 60 asl. There are deep and branching gorges in sandy soil sloping S–SW up to 60 m asl in the Kuukso village, 9.8 km NW of Raavuori. After apparently stormy flooding, the returning water eroded the gorges. The gorges can be seen in the hillshade derivations of the laser mapping (MapSite 2022–2024).

The recorded observations of the wrecks of five ancient ships in the inland region were compiled (Ahokas 2003) and are positioned on the map (Fig. 1.1), appearing evidently at current altitudes ≤40 m asl. The prehistoric, wooden figurehead of a vessel (Erä-Esko 1958) found in a swamp at 96.5 m asl in Rovaniemi, North Finland, was dated with geological bases to the beginning of the Litorina era (Hyyppä 1958), and its wood was later radiocarbon dated to 9.01–8.20 cal ka BP (Carpelan 1974: Hel-168). Ethnographer S. Paulaharju (1932: p. 147) recorded a finding of an ancient ship in an inland swamp in Härmä, Ostro-Bothnia, W Finland. On the

Ostro-Bothnian coast, the annual land upheaval is fastest in Finland, up to 9 mm in the early half of the 1900s (Kääriäinen 1963). Known prehistoric rock art south of Ss1 is limited, compared with that between Ss1 and Ss2, where they appear between 83 and 63 m asl (Kivikäs 1995; Miettinen 2000), corresponding to Yoldia and Ancylus shore levels. South of Ss1, rock art appears between Yoldia and Litorina shore levels.

4.6 Supercooling and Ice Blockages in Rivers, and High Precipitation Resulted in Floods

The region associated with the Kymi River south of Ss2 is a depression, the Kymenlaakso, 'Kymi Dale', which has shown faster land upheaval than the regions on both sides (Fig. 1.1, Kääriäinen 1963). The glaciofluvial eskers of practically N–E direction have significantly smaller tilting per km of the sand–clay contact levels in the strip where they meet the Kymi river (Fig. 1.1; Ahokas 2003). This has resulted from the faster land upheaval in the Kymi Dale, but has apparently also directed the run of the Kymi-to-be after the Ice Age to the existing dale. Since Kääriäinen's publication (1963), the upheaval rate has slightly slowed (Kakkuri 1987). After the deglaciation, the upheaval rate in SW Finland was considered to have been several times that in the present time (Eronen et al. 2001).

Ice blockage by freezing to the bottom of the "Vuoksi rapids", probably meaning the Imatra rapids, was remembered in the Finnish oral tradition (Ganander 1789). Likewise, the ancient Kymi has been blocked by driving ice and freezing to the bottom, which caused serious floods at times (Blomqvist 1911; Kokkonen 1926). The radiocarbon-dated ages (Table 3.2) and toponyms reminding of wetlands (Table 4.1) also suggest that wide flood lagoons downstream of the Kymi occurred and were the reason for part of the swamps. The local history since the sixteenth century has known major floods of the Kymi (Bergström 1703; Alfthan 1908; Blomqvist 1911; Halme 1969; Oksanen 1981; Seppovaara 1988). Many elevations in the terrain staying above the dangerous floods, earned a toponym having the meaning of an island (Table 4.2).

Approximately by 7.5 cal ka BP, the summer temperature had declined in the northernmost Europe from its maximum in the Holocene (Väliranta et al. 2015). North of the catchment of the Kymi, in the Tsuolbmajavri Lake area, North Finland, high annual precipitation was inferred to have prevailed between 9.0 and 7.0 ka cal BP (Seppä and Birks 2001). The luminescence-dated generation of the Koria sand plains at 7.61 ± 0.46 ka, which is the earliest dating associated with the outburst of the proto-Kymi through the Ss1 ridge, probably coincided with the rainy era in its catchment area.

The channel of the Kymi is younger than the calcium carbonate concretions on its bottom. The Kymi has eroded its channel in the hard varved clay, and its bottom is covered with minimal soft clay sediment as encountered by the diver (Hemmilä

Table 4.1 Toponyms describing wet areas, most of which were apt to be subjected to floods south and west of the Kymi, downstream of the First Salpausselkä ridge in the northern and central parts of the historical Elimäki municipality, before the effective draining with channels and ditches. The places were evidently above the Litorina maximum

In modern orthography	Extracted from old maps and texts, with some radiocarbon (^{14}C) ages
aro = 'grassy swamp'	
Arola	(Anon. 1777; Oksanen 1985: p. 198)
Aropelto	Aropeldo (Ottelin 1793)
Lammas**aro**	Lammasaronkallio (Frisius 1752), Lammastenaro (Frisius 1765), Lammasaro (Barck and Riddelin 1835)
Lehvistö**aro**[a]	Läwisto aro (Westermark 1774–1776)
Martin**aro**	An evident past quagmire, a wartime shelter of the escaped Nappa people (Pakkala 1902)
Näsin**aro**	(R. Hemmilä, pers. commun. 2024), same as "paistivan oja" (<1788, Fig. 1.2)
Niiniojan**aro**	Nijnojan aro (Frisius 1758)
Nisu**aro**[a]	(Pakkala 1902)
Puusuonkankaan**aro**	Pussukanga(ste)tenaro (Törn 1753), Puβukangan aro, Pu(o)sukangan aro (Frisius 1758)
Ruohois**aro**[a]	Ruohoisen aron perä (Erichzsonn 1639; Israellsonn 1640)
Suvenojan**aro**	Suvenojan aro (Frisius 1752), Sufvenojan aro (Frisius 1758)
[b]**kor = 'swampy'**	
Korpi	Korpiniemexet, Korpipeldo (Ottelin 1793)
Korp'aita	Korpaidanaho (Ottelin 1793)
Kortila, **Kor**ja, **Kor**ia	Kårtil(l)a (Anon. 1777; before 1788 in Oksanen 1985: p. 198), Korja (Barck and Unonius 1837; Oksanen 1985: p. 393)
Korvenkylä[a]	Korpis by (Anon. 1777), Korvenkylä (Anon. 1875)
Paha**korpi**	Pahakorven nitut (Ottelin 1793)
Peräkorpi	Peräkorven nitu (Frisius 1758)
Pukki**kor**pi	Bockkorfvenahot (Ottelin 1793)
Ratas**kor**pi	Ratas Corven Nitu (Anon. 1777)
Rita**kor**pi	(Frisius 1758)
Seppälän**kor**pi	Seppälänkorfvenaho (Ottelin 1793)
ropakko = 'muddy place'	
Ropakko[a]	Råpako (Westermark 1774–1776)
Ropakkoniitty, -pellot	Ropackonitu, Ropackopellot (Ottelin 1793)
suo = 'swamp', 'mire'	
Haapa**suo**	Hapasuo (Lewong 1773)
Haapa**suo**[a]	(Westermark 1774–1776), Hapa Suo (Anon. 1777)
Hevos**suo**[a]	Häffwo soon kifwi (Ericβon <1667)
Hevos**suo**	Hävos suo (Anon. 1777), Hevossuo (Luukkonen and Luostarinen 1913–1915)
Holstin**suo**[a]	Holstin suo (Westermark 1774–1776)
Järven**suo**[a]	Järven suo (Westermark 1774–1776)

(continued)

Table 4.1 (continued)

In modern orthography	Extracted from old maps and texts, with some radiocarbon (^{14}C) ages
Kangasojan**suo**	Kangas åijanså (Anon. 1653)
Kapulasillan**suo**	Also called Röntynsuo (Vainio 1918)
Karju**suo**	(Luukkonen and Luostarinen 1913–1915)
Kaukainen**suo**	Kaug. så sarj, Kauken så sarj (Hinrichson 1635)
Kauka**suo**[a]	Village name in Anjala. Kaukasuo-1 ^{14}C dating, 1058–727 cal a BP (Table 3.2)
Kengäsojan**suo**	(Ottelin 1793). Kengäsaho-1 ^{14}C dating, 2357–1888 cal a BP (Table 3.2)
Kuren**suo**[a]	Kurren sou (Bergström 1703), Kurren suo (Westermark 1774–1776; Anon. 1777), Kurensuo (Pakkala 1902)
Lakia**suo**[a]	In Anjala (Westermark 1773, 1774–1776)
Lakia**suo**	In Elimäki (Ottelin 1793)
Lassilan**suo**[a]	Lassilan Suo (Anon. 1777)
Mieliös**suo**	Melinsuo (Kurck 1646), Melionsuo (Ottelin 1793)
Mustan**suo**	Musta Mores Suo (Anon. 1777; Luukkonen and Luostarinen 1913–1915)
Nälkä**suo**[a]	Nälckäsuo (Westermark 1774–1776)
Napa**suo**	Modern maps
Porras**suo**	Porasuonpeldo (Ottelin 1793), Porrassuo (Anon. 1875)
Potikko**suo**	Potikosuo (Frisius 1752; Törn 1753)
Puolitien**suo**	Puolitensuonsarenpä (Frisius 1752)
Puu**suo**	Pussukanga(ste)naro (Törn 1753), Puβukangan aro, Pu(o)sukangan aro (Frisius 1758), Pussunkangas (Barck and Riddelin 1835)
Pyöräs**suo**[a]	Pyörän suo (Westermark 1774–1776), Pyöres-Suo (Anon. 1777)
Raja**suo**[a]	Rajasuonmäki (Westermark 1774–1776)
Rasikeson**suo**	Raskkesonsuo (Frisius 1758)
Retkus**suo**[a]	(Westermark 1774–1776), Rättkus Suo (Anon. 1777)
Riipin**suo**[a]	Riipin suo (Pakkala 1902)
Rita**suo**	Rita suo (Ottelin and Kihlström 1792–1810)
Ruohoisenaron**suo**	Rohoisen aron suo (Israellssonn 1640)
Ruoko**suo**	Ruokosuonaho (Ottelin 1793)
Soidin**suo**[a]	Soininsuo (Westermark 1774–1776)
Sorsalammin**suo**[a]	Sorsanlammin suo (Westermark 1774–1776), Suorsa lammin suo (Anon. 1777)
Suokonpelto	Suokonpeldo (Ottelin 1793)
Suolanoja	Suolan oja (Munselius 1666)
Suonpää	Suonpä (Ottelin and Kihlström 1792–1810)
Suurpalon**suo**[a]	(Westermark 1774–1776)
Suur**suo**[a]	(Westermark 1774–1776; Westermark and Barck 1774–1848)
Suvenojanperän**suo**	Suwenojanaronperensuo, Suwenojanaronojansuo (Frisius 1758), Suviojan pären suo (Anon. 1777)
Sydän**suo**[a]	Suvänsuo, Sydänsuo (Westermark 1774–1776), Sydensuo (Westermark and Barck 1774–1848)
Talvitien**suo**[a]	Talwitiensuo (Westermark 1774–1776)

(continued)

Table 4.1 (continued)

In modern orthography	Extracted from old maps and texts, with some radiocarbon (^{14}C) ages
Vehka**suo**	Vehka Suon nitu (Anon. 1777)
Veitsaarennon**suo**[a]	Weitsaarendonsuo (Westermark 1774–1776)
Yrtti**suo**[a]	Ÿrttisuo nijttu (Erichzsonn 1638), Ÿrttisuon nijttu (Israellsonn 1640)
suolento, sulento = 'paludification'	
Kivi**sulento**[a]	Kifwi sulendo (Bergström 1703), Kivisulendo (Westermark 1773, 1774–1776)
Lyly**suolento**	Lÿlÿ suolendo (Anon. 1481)
Sulento[a]	Suolento, abbreviated to the modern Sulento
v**ataja, vesi = 'watery'**	
Vatajanaho	(Ottelin 1793). Simolanaho-1 ^{14}C dating, 2321–1940 cal a BP (Table 3.2)
Vesistenaho	Wesistenaho (Ottelin 1793)

[a] In the territory of the Anjala municipality, earlier the eastern part of Elimäki (Hausen 1920; Oksanen 1981)

[b] Additional **kor**- toponyms nearby the Kymi, **Kor**jala = a suburb beyond the Kymi at Koria (Fig. 1.2), and **Kor**järvi, a swamp-lake 3 km east of the Kymi in the former Sippola municipality

Table 4.2 Toponyms describing areas usually staying above the floods of the Kymi, downstream of the First Salpausselkä ridge in the northern and central parts of the historical Elimäki municipality, before the effective draining with channels and ditches

In modern orthography	Extracted from old maps and texts
mannari = 'elevated terrain'	
Mannari	In the field west of the Tolkkila house, Värälä village (Punttila 1960), and its detection with the laser-scanning (MapSite 2022–2024)
Mannari	The hill of wartime shelter of the escaped Vilppula villagers (Pakkala 1902)
napa = 'hill-top'	
Napa, Nappa	Napa (Anon. 1658, 1842; Frisius 1765), Napapuolenvainio (Luukkonen 1906–1907), Napasuo (modern maps), Nappa (Törn 1753; Frisius 1758)
Napamäki	(Vainio 1918)
Paavon**napa**	Pavon napa (J. O. Selandro 1650 in Frisius 1752), pavonnappa (Frisius 1758)
saari, saares, saare, saarento = 'island'	
Haan**saari**	Hansari (Starck and Bonsdorf 1828–1831)
Haapa**saari**	Hapasarj (Pedherson 1568), Hapsari (Bergström 1703; Westermark 1773, 1774–1776)
Haasia**saari**	Haβia Sari (Frisius 1765)
Hongan**saari**	(Vainio 1918)
Joensuun**saari**[a]	(Westermark 1774–1776)
Kaali**saari**[a]	Kali sari (Bergström 1703)

(continued)

Table 4.2 (continued)

In modern orthography	Extracted from old maps and texts
Kangas**saari**	Kangansaaren nijtu (Kurck 1646), Kangassaari (Meurman 1908–1911), later called Rajasaari
Karja**saari**[a]	Carjasaari (Westermark 1774–1776), Karjasaari (Westermark and Barck 1774–1848)
Kaski**saari**[a]	Casksaari (Westermark 1774–1776), Kaskisaari (Westermark and Barck 1774–1848)
Kaukaisensuon**saari**	Kaug. så sarj, Kauken så sarj (Hinrichson 1635)
Kiviportin**saarento**[a]	Kiviportinsaarendo (Westermark 1774–1776)
Kota**saari**	(Anon. 1466)
Lato**saari**	LatuSari (Lewong 1772)
Lautak**saari**	A rock in Vilppula village
Maja**saari**[a]	(Westermark 1774–1776)
Mänty**saari**[a]	Mändysari (Westermark 1774–1776)
Mänty**saari**	Tall holma (Anon. 1777) in Swedish
Marjo**saari**[a]	Mariosaari, Marjosa(a)ri (Westermark 1774–1776; Westermark and Barck 1774–1848)
Matara**saari**[a]	Matarasa(a)ri (Bergström 1703; Westermark 1773), Matair saari (Westermark 1774–1776) in the Lakiasuo[a] swamp
Merta**saari**[a]	Merttsari (Erichβon 1667)
Palo**saari**	Palosaaren luchta (Sigfwersson 1544), Palon Sari (Anon. 1777)
Perä**saarento** or Perä**saari**[a]	Pärä sarendo (Bergström 1703), Pärä saari (Westermark 1774–1776)
Puolitiensuon**saari**	Puolitensuonsarenpä (Frisius 1752), Puolitiensuonsarenpä (Frisius 1758)
Raanojan**saari**	Rajåiansari (Israellsonn 1640), Raianåjansari (Anon. 1679)
Rataskorven**saari**	Ratas Corven sari (Anon. 1777)
Rita**saarento**[a]	Ritasarendo (Bergström 1703)
Saarestenkallio or **Saare**nkallio[a]	Sarresten Kallio (Anon. 1625), Saren callio (Westermark 1774–1776; Anon. 1777)
Savi**saari**	Sawisari (Frisius 1752), Safwisari (Frisius 1765), Savisaari (Barck and Riddelin 1835)
Suuri**saari**[a]	Suri Sari (Anon. 1777), Suursaarenmäki (Anon. 1875)
Suuri**saari**	Suurisaarenmäki (Luukkonen and Luostarinen 1913–1915)
Sysi**saarento**[a]	Sÿssarendo (Westermark 1773)
Urt**saarento** or Urtin**saari**[a,b]	(A. Kujala in Kokkonen 1927; Oksanen 1981)
Väli**saari**	Veli Sari (Anon. 1777)
Veit**saarento**[a]	Weitsaarendonsuo (Westermark 1774–1776)
soppi = 'eminence'	
Sopenpalsta	(Luukkonen and Luostarinen 1913–1915)

[a] In the territory of the Anjala municipality, earlier the eastern part of Elimäki (Hausen 1920; Oksanen 1981)

[b] The archaic Finnish *urtti* (genitive *urtin*) denotes 'deep, narrow watercourse in rapids' (Itkonen et al. 1975) and suggests here a past channel of the Kymi

2017). Ca-rich carbonate concretions frequently occur on the bottom of the Kymi several kilometers downstream from the Nappa area in Koria (Fig. 1.2), where the concretions emerge from the eroded clay to the river bottom (Hemmilä 2018: p. 96). Such concretions were elsewhere in southern Finland dated between 10 and 9 ka BP (Eriksson 1987), i.e., the Ancylus era. An intact Ca carbonate cobble from the coast of the Gulf of Finland was radiocarbon dated to 10.1–9.7 cal ka BP (Beta-441226; Ahokas 2017: Table 26).

The water current in the Kymi varies between 96 and 520 m^3/s (Hemmilä 2017: p. 139), and reached 856 m^3/s in the Voikkaa rapids in the rainy and high-snow epoch of 1898 and 1899, which caused a 3-m surplus over the typical flood maxima around the river (Alfthan 1908: p. 51). Low temperature periods causing supercooling of water (see the Sect. 1.2) and blockages of the Kymi at Koria might have occurred and coincided with rainy seasons. The profiles of Simolanaho-1 and Kengäsaho-1 sample sites also suggest the occurrence of increased precipitation with low evaporation about 2300–1900 cal a BP (Table 3.2). The climate in NW Europe had turned from relatively warm and continental to oceanic between *c.* 2800–2710 cal a BP (van Geel et al. 1996). Between 2857 and 2825 cal a BP, the mineral matter accumulation also increased in a studied lake in central Finland (Tiljander et al. 2003). The Late Antique Little Ice Age has been maintained in the Norse tradition as *fimbulvinteren*, or exceptionally cold seasons, which were dated by different means to 1414 cal a BP and some subsequent years (Oinonen et al. 2020; Anagnostou et al. 2025). The Suppamäki-1 radiocarbon age, 1529–1309 cal a BP (40 m asl, Su-3078; Table 3.2) could fit to the Late Antique Little Ice Age as a putative ice blockage and subsequent flooding period of the Kymi.

Charcoal from fluvial sand of the Pärnäjoki river ('Linden-river') was calibrated to the calendar years 1397–1656 (Table 3.2). The apparent wood formation of the charcoal overlaps the time when the fishing court seats dealing with a share of fishing rights in 1460 and 1461 (Anon. 1460, 1461) on the Pärnäjoki river, showing the water currents then to have maintained meaningful stocks of fish, perhaps thanks to the floods. The court determined the fishing boundary between Mr. Anders Timberman and the Raussila village people. In 1461, the Raussila part was defined as the person Mr. Peter Raisian. The raw point was at "Karjanoia" ('Livestock-ditch'; Anon. 1460), or "Ojansilda" ('Ditch-bridge'; Anon. 1461), for which the abbreviated name Oja ('Ditch') appeared in the twentieth century (Pfäffli and Hellman 1918), being a gorge and ditch with pasture. The raw point is east of a hill formely called "Mäurän kiwenhariu" (Anon. 1460), "Meuranpesänharjus" (Anon. 1461), later "Meurenmäki" (Anon. 1777), and today, Pieni Mäyrämäki ('the Little Badger-hill'), where, on the western Pärnäjoki shore at the raw point, there are two big boulders lying 10.6 m apart, which confirmed the particular raw point: "*ther äre 2 store stenar*" ('there are 2 big stones'; Anon 1461). In modern times, the boundary of municipalities (Anon. 1842), villages, and farms lay midway between these big erratic boulders. This raw point is 1.1 km downstream of the site of collecting the dated Pärnäjoki-1 charcoal (Table 3.2).

The calibrated Pärnäjoki-1 dating for a flood and putative ice blockages of the Kymi in 1397–1656 fits to the Little Ice Age in Fennoscandia at about 1600 AD

Fig. 4.2 Ilmajärvi was an old name of this shallow, now tillable lake basin. Forms of the old written name were, e.g., "Jlmeiärffue" in 1559 and "Jlmi Jerffui" in 1593 (Hausen 1920). The Finnish *ilma* appellative refers to 'an aptness of flooding'. The modern official name of the lake and the current field is Elimäenjärvi ('the lake of Elimäki'), but locally called only Järvi ('the Lake'). The lake basin was drained and taken into cultivation in the mid-twentieth century (Oksanen 1985). The conical water tower in the background forest at 2.15 km from the photographer stands on a rock 53 m asl. The Tyynelä rock painting (the position R-T in Fig. 1.1 and the motif in Fig. 4.1b) occurs between 42 and 43 m asl on the upright face of the rock, 28 m east of the water tower, pointed with the blue arrow. The painting was probably made by standing on the ice of the X sea, dated in this book to have existed between 6.9 and 6.7 cal ka BP. The deepest measured organic-rich sediment of the lake basin exceeds 32 m (Anon. 1980). Flood water has streamed into the lake, retarded the speed, and unloaded sediments. Water of the Kymi river from behind 14 km NNE, the Ahkoja area, has flooded to this basin, in the historical era via a channel, the Pärnäjoki river. The flooding is frequent in the spring after the snow melting, as illustrated in Kaipainen (2022), but it this time occurred after summer showers and rose to about 29 m asl. The pumping is able to lower the water by about 0.2 m a week (Kaipainen 2022), when the flood covers the extent of the basin. Photo: July 19, 2025

(Briffa et al. 1990; Moberg et al. 2005). The variable weather during the Little Ice Age at times caused high precipitation and floods in Europe (Fagan 2000; Wanner et al. 2022). At a distance of 350–400 km from Kymi Dale, in the Lake Ilmen at Novgorod, floods were indicated to have occurred in 1436, 1437, 1455, 1456, 1468, 1495, 1518, 1540, 1544 and 1555 AD (Akiander 1849), i.e., in the Little Ice Age period. The name Ilmen (in Russian Ѝльмень, old-Russian Íl̆mĕrь) is derived from the Finnish Ilmajärvi (Vasmer 1953), and appeared as Ilmajervi in Teitti (1555–1556), Ilma in Jenkinson's map (1579), and Ilmjärv in Estonian (Vasmer 1953). The Finnish *ilma-* has the archaic geographical or hydrological sense of 'apt to flooding' (Fig. 4.2), and was also loaned to the meaning of 'a small lake left after high water in the Volga' in local Russian (Vasmer 1953). In Lake Ilmen, the water surface area is known to vary 3.6 times between 620 and 2230 km^2 (Kokorev 2003).

4.7 The Order of the Outburst of the Vuoksi River and the Ladoga Transgression

The large Vuoksi river and the small Hiitolanjoki river wore their channels through the Ss1, respectively, 123 and 164 km ENE of the Kymi Ss1 channel. For the Vuoksi outburst, the dating by Saarnisto (1970) is routinely taken as the age (6.18–5.66 cal ka BP) based on wood pieces in the sediment 63 m asl (Linnansuo, Hel-8). The ignored, almost contemporary radiocarbon dating of 9.56–8.60 cal ka BP (Imatra, Su-49) from the same Linnansuo swamp, 64.65 m asl (Hyyppä et al. 1969), should also be considered for the dating of the Vuoksi outburst. The Imatra (Su-49) sample comprised *Bryales* peat containing remnants of deciduous trees. The water streaming near the Hel-8 sampling site has apparently worn away the older and 1.65 m higher layers and relocated younger material there at the 63 m asl level. The Vuoksi rapids were remembered to have been blocked to the bottom by freezing (Ganander 1789; see supercooling in Sect. 1.2). Ends of the ice blockages can have liberated great volumes of accumulated water to flow, erode, and relocate the soil.

The water from the Vuoksi outburst may have partly flowed into the Gulf of Finland, e.g., via the route still known in the historical era (Teitti 1555–1556), and partly into the Ladoga, where the outburst was apparently mainly responsible for a Ladoga transgression.

The radiocarbon dating laboratory at the Geological Survey of Finland was started in 1960, and its methodology was shortly described in Hyyppä et al. (1962). The Ladoga transgression was for the first time radiocarbon dated by Hyyppä et al. (1964) to 8.32–7.62 cal ka BP from wood (Metsäpirtti, Su-28) sampled in 1937 from the upper part of the peat section under silt and fine sand on the shore of the Viisjoki river, on the western coast of the Ladoga. The Viisjoki site in the Metsäpirtti municipality, the territory of Finland back then, was illustrated and described by Hyyppä (1942). From the shore of the same Viisjoki river, three additional radiocarbon ages for the Ladoga transgression are given in Table 4.3. Viisjoki appears there under the Russified name Вюон or Vyon, introduced by the Soviet Union (Znamenskaya et al. 1970; Liiva et al. 1978). Two additional ages from samples of two other sites of coastal rivers flowing to the Ladoga, and one bottom sediment dating of the northern Ladoga, 121 m below the surface, are given (Table 4.3). The overall range of these seven ages is 8.37–6.56 cal ka BP. This whole range of the transgression ages ranks the older age of 9.56–8.60 cal ka BP (Imatra, Su-49 by Hyyppä et al. 1969) to be the probable age of the outburst of the Vuoksi to the Ladoga capable of explaining the Ladoga transgression, instead of the more recent age of 6.18–5.66 cal ka BP presented to date the Vuoksi outburst by Saarnisto (1970). The 6.18–5.66 cal ka BP age would suggest the outburst of the Vuoksi to

Table 4.3 Radiocarbon (^{14}C) dating of samples from the coast of the Ladoga buried under mineral deposits, plus one from the bottom sediments. They are thought to present the Ladoga transgression contemporaneous with the Atlantic period

Site [on the coast of the Ladoga]	Altitude asl (m)	Sample name [sampling year]	Depth (m)	Dated material	Laboratory number	^{14}C age a BP	Calibrated age ka BP	References
Viisjoki [W]	13.9	Metsäpirtti [1937]	≥2	Burried peat	Su-29	7100 ±170	8.32–7.62	Hyyppä (1942), Hyyppä et al. (1964)
Viisjoki (Вюон) [W]	14…15	Вюон [≤1967]	≈4.4	Burried wood, aspen	LE-561	5980 ±100	7.16–6.56	Znamenskaya et al. (1970)
Viisjoki (Vyun) [W]	14	Vyun I a [1972]	5.8 −5.9	Buried wood, alder	Ta-488	7215 ±160	8.37–7.74	Liiva et al. (1978)
Viisjoki (Vyun) [W]	14	Vyun II a [1972]	≈5	Burried wood	Ta-513	6570 ±70	7.58–7.33	Liiva et al. (1978)
Paksujoki [S]	<15	Паша [1958]	7.52 −7.82	Buried peat	LE-333	6200 ±160	7.43–6.75	Lisitsyna (1961), Znamenskaya et al. (1970)
Ust'-Rybezhno I at Paksujoki [S]	<15	Усть-Рыбежно I [1958]	0.8 −1.06	Buried charcoal[a]	LE-405	6380 ±220	7.67–6.76	Lisitsyna (1961), Kvasov and Nazarenko (1970), Znamenskaya et al. (1970)
61° 33′ 25″N 31° 20′ 7″E, the northern Ladoga	Below the surface: 121 m[b]	1097 [The sampling year not stated[c]]	No data	Bottom silt	LU-2817	6660 ±220	8.01–7.02	Arslanov et al. (1996)

[a] Wood charcoal of the Neolithic cultural layer buried under sand. This Neolithic society moved to a new place (Усть-Рыбежно II) 5 m higher and about 1 km away (Lisitsyna 1961), evidently forced by the Ladoga transgression

b The surface of the Ladoga was 5.2 m asl (Kare 1929)

[c] Sampling probably after the Chernobyl fallout, the intensity of which in the northern Ladoga is not known. The fallout radiation tends to increase the ^{14}C proportion formed from nitrogen-containing material, making the result appear younger than it is

have occurred after the onset of the Ladoga transgression, though the Vuoksi outburst was probably the principal water source of the Ladoga transgression. The modern Vuoksi makes about 29% of the total drainage to the Ladoga (Kuusisto 1982). Hyyppä (1942) concluded that this particular Ladoga transgression had occurred during the Litorina era, which proved to be coherent with the radiocarbon age (Hyyppä et al. 1964). The Ladoga surface is generally thought to have been 1 or 2 m above the past Litorina Sea.

The millennia-old Finnish oral tradition remembered the Ladoga transgression by the local inhabitants on the Mantsinsaari island on the Ladoga until the late 1800s. Then Ailio (1897: p. 38) published the oral tradition that the Ladoga 'was dammed' (*oli salvassa*) in the past. The Finnish *salpa*, 'dam', 'blocking', 'bar' (in the singular inessive case, *salvassa*), also appears as part of the geographic names of eskers, like Salpausselkä, 'damming ridge' (Fig. 1.1). The Finnish archaeologist, Dr. Julius Ailio ((born in 1872, death in 1933), during his research carrier, erroneously thought the transgression in the Ladoga to have been caused by the Litorina Sea transgression. The dam, *salpa* of the Ladoga ended, and the Ladoga transgression declined with the outlet crossing the Karelian Isthmus to the Gulf of Finland through the temporary Heinjoki strait (Ailio 1915) starting *c.* 4.10 cal ka BP (Ludikova et al. 2024), and finally with the persisting outlet river, the Neva, centuries later (Saarnisto and Grönlund 1996; Ludikova et al. 2024). The Finnish *neva* denotes 'swamp' and also i.a. 'swampy river' (Toivonen et al. 1958), and is a common toponym in Finland: the name database of Finland (Nimisampo 2024) finds 31,250 Neva placenames, or placenames with *neva* as the appellative.

The unequal post-glacial isostatic upheaval speeds, slower on the southeastern edge in the territory of Finland, have inclined and incline this land towards southeast (Kääriäinen 1963; Okko 1964). The 9.56–8.60 cal ka BP age (Hyyppä et al. 1969) for the outburst of the Vuoksi in the southeast is in accordance with the inclination of the land culminating in the water at the Vuoksi prior to the proto-Kymi having its outburst at 7.61 ± 0.46 ka as dated now. Water level tilting of centimeters towards SE in two inland lakes in the catchment region of the Vuoksi has been measured over 50 years, since 1912 (Sirén 1963).

4.8 ^{14}C from Nitrogen, ^{14}N Containing Substances Under Environmental Irradiation

Besides the unavoidable cosmic radiation (Kitagawa and van der Plicht 1998), the radioactive radiation from soil and anthropogenic fallouts can increase the ^{14}C isotope from ^{14}N and make determined radiocarbon ages younger than they truly are in samples containing nitrogen. The proportion of the ^{14}C isotope is typically only 0.0000000001% (Beta Analytic 2018). The Kymi Dale region is on rapakivi granite rock (see Sect. 1.1), which is fairly rich in U and Th. ^{14}C increases in U and Th ores (Jull et al. 1987; Bonetti et al. 1999), and with endogenous neutron radiation in the granite of Stripa in Central Sweden (Andrews et al. 1989). The U and Th contents from the aeroradiometric measurements (GTK 1992–2001) are given in Table 3.2, because radiation measurements may provide a tool to increase accuracy in calculating true radiocarbon ages in samples containing nitrogen.

Additionally, in 1986, the weather conditions brought plenty of radioactivity to Finland from the explosion and fire of the Chernobyl nuclear power plant in the Soviet Union, as early as 12 h after the explosion (Pöllänen et al. 1997). In the region of the study, particle emissions also arrived later from the Russian Sosnovyj Bor nuclear power plant. The Elimäki municipality was among the highest polluted areas in Finland from Chernobyl in terms of ^{137}Cs radiation calibrated on Oct. 1, 1986, the surface activity being up to 100 kBq/m^2 (Arvela et al. 1987, 1990), and that of ^{103}Ru surface activity in April 1986 being up to 40 kBq/m^2 (Arvela et al. 1990) in addition to many other radioactive isotopes. At most, the hot particles from the Chernobyl fallout on land reached a density of over 10,000 per m^2 (Toivonen et al. 1988).

At the moment, we lack ways to correct radiocarbon ages, possibly biased by man-made radiation on the sample. The putative increase in ^{14}C is also dependent on the nitrogen content of the sample. Identical materials sampled before and after being subjected to, e.g., the Chernobyl fallout, would be interesting to compare.

4.9 The Koria Sand Plains Were Cultivated with *kytö* Burning

The dated charcoal layer on the Koria sand plains, calibrated to the seventeenth or eighteenth century (Figs. 1.2, 3.6a, b, Table 3.3), shows fire cultivation to have been practiced here. In the *kytö* technique for fertilization and controlling weeds and pests, soil was burned or heated with a flameless fire with extraneous wood fetched annually or at some years' intervals (Ahokas 2012). Winter or spring floods of the Kymi probably made the seasonal soil moisture suitable in the fine Koria sand with a moisture reservoir in the clay underneath (Fig. 3.6a, b), e.g., for turnip cultivation in the summer. In the concerned village, still in the 1820s, there was a meadow parcel called Kytöniitty ("Kytanitu"; Starck and Bonsdorf 1828–1831). Some other earlier toponyms in the northern Kymi Dale are Kytöniitty ("Kytanijt"; Larβon 1650, or "Kÿten nijttu"; Larβonn 1653), Ky'önlähiaho ("Kÿönäär aho"; Larβonn 1653) and Kytötarhamaa ("Kyttötarhama"; Frisius 1752). The Kytö parcels showed that the *kytö* burning technique was traditional in this area, even a *kytö*-burner peasant, Matti (*Kÿttarn Bonden Matz*), was termed in an old document on the eastern side of the Kymi, in the Mämmälä village (Anon. 1659). In the *kytö* burning, the danger of wildfires was less than in the slash-and-burn (*kaski* in Finnish), which also needed more people to work. King Gustaf the First tried to forbid the *kytö* burning in 1559 (Ahokas 2012: p. 25). Long-term *kytö* burning ultimately made the soil infertile.

The past land surveyors and landlords were usually Swedish speakers in the East Land (Finland-to-be) of the Kingdom of Sweden. The Swedish speech tends to extend the continuant of consonants or to geminate them. Therefore, there occurs for the Finnish *kytö*, the Swedish forms *kyttö*, and its derivative *kÿttar* (see above). Likewise, for the Kymi river, the Swedish has introduced the geminated Kymmi (Frisius 1765). The official Swedish name of the Kymi has the geminate like Kymmenæ (Ioansson 1380), probably adopted from the Finnish essive case *kymenä* ('as a river', 'as the Kymi'). The modern Swedish orthography is Kymmene. The nearby village name Nappa has been geminated by the Swedish language from the Finnish word *napa*, in the geographical sense meaning 'a hill-top' or an elevation not reached by the floods. The original form Napa, instead of Nappa, can be found in Swedish texts (Anon. 1658; Frisius 1765), and in Finnish texts, Napapuolenvainio parcel name (Luukkonen 1906–1907) and Napasuo toponym in modern maps (Table 4.2).

References

Ahokas H (1992) Endospermal protein diversity, as revealed by electrophoresis and Ig binding on blotted membranes, in Leymus arenarius from non-littoral, roadside niches. Ann Bot Fenn 29:295–304

Ahokas H (2003) Major discharge and floods of the Kymi river (Kymijoki) over the Salpausselkä I esker to coastal South Finland traced by ^{14}C datings, unequal land upheaval, oral tradition and onomastics. Kave, Helsinki

Ahokas H (2006) Tähkätädykkeen, ruoholaukan ja keltamaitteen reliktiesiintymistä sisämaan muinaisrannoilla. Summary: inland occurrences of Veronica spicata, Allium schoenoprasum and Lotus corniculatus at ancient Baltic Sea shore levels. Lutukka 22:45–57

Ahokas H (2012) Crop evolution under fire. Interdiscip Biol Agric Linguist Antiq 5:1–194

Ahokas H (2015) Aikaisemmin tunnistamattomia muinaistaidekuvia, hieroglyfi, petroglyfejä ja kalliomaalauksia Kaakkois-Suomesta. Abstract: Recently found hieroglyph, other petroglyphs and rock paintings in South-East Finland and about the helsink appellative in toponyms. Interdiscip Biol Agric Linguist Antiq 6:1–63

Ahokas H (2017) Aiemmin tunnistamattomia meteoriittitörmäyksiä Etelä-Suomeen. Abstract: Previously unidentified meteorite impacts in South Finland. Interdiscip Biol Agric Linguist Antiq 7:1–75, 28 tables, 182 figs

Ahokas H (2018) Ruoholaukan uusi löytöpaikka muinaisrannalla Kymenlaaksossa ja mustan biofilmin merkitys. Summary: Allium schoenoprasum on the prehistoric lake shore at a recently discovered inland site in SE Finland. Lutukka 34:104–111. https://tietopankki.luomus.fi/lehdet-ja-kirjat/lutukka/

Ailio J (1897) Über Strandbildungen des Litorinameeres auf der Insel Mantsinsaari. Fennia 14(2):1–43, map

Ailio J (1915) Die geographische Entwicklung des Ladogasees in postglazialer Zeit und ihre Beziehung zur steinszeitlichen Besiedelung. Fennia 38(3):1–107, 123–159, illustration, 2 maps

Akiander M (1849) Utdrag ur ryska annaler. Suomi 8:1–284

Alfthan M (1908) Bidrag till kännedom af Kymmene älf. Résumé: Contribution à l'étude du Kymmene. Fennia 27(3):1–96, appendices

Almén A-K, Glippa O, Pettersson H, Alenius P, Engström-Öst J (2017) Changes in wintertime pH and hydrography of the Gulf of Finland (Baltic Sea) with focus on depth layers. Environ Monit Assess 189:147. https://doi.org/10.1007/s10661-017-5840-7

Anagnostou E, Linderholm J, Lidén K (2025) The AD 536/540 climate event in Sweden—a review. Boreas 54:1–13. https://doi.org/10.1111/bor.12672
Ancient shorelines (2013) Special data products. The Geological Survey of Finland, Espoo
Andrews JN, Davis SN, Fabryka-Martin J, Fontes J-C, Lehmann BE, Loosli HH, Michelot J-L, Moser H, Smith B, Wolf M (1989) The in situ production of radioisotopes in rock matrices with particular reference to the Stripa granite. Geochim Cosmochim Acta 53:1803–1815
Anonymous (1460) Anno Domini 1460. Feria quinta proxima post Festum beatæ Margaretæ virginis. Tå war fisk tingh i Elemä. In: Hausen R (ed) (1881–1883) Bidrag till Finlands historia, vol 1, p 120
Anonymous (1461) Anno Domini CIↃ C IↃ LXI Vigilia S. Olai. Tå war fisketingh i Elimä sochnn. In: Hausen R (ed) (1881–1883) Bidrag till Finlands historia, vol 1, pp 120–121
Anonymous (1466) Feria 3[a] proxima ante Festum Margaretæ virginis. Tå war sommar tingh i Elimä. In: Hausen R (ed) (1881–1883) Bidrag till Finlands historia, vol 1, p 121
Anonymous (1481) Domboken för sydöstra Tavastland 1443–1502. Microfilm PR1A. The National Archives of Finland, Helsinki. Typographically in: Hausen R (ed) (1881–1883) Anno Domini 1481. Feria sexta post festum S. Olai, då war sommar tingh i Tennilä sochn i Pollwola by. Bidrag till Finlands historia, vol 1, p 33
Anonymous (1625) Dhen 8 octobris åhr 1625. Wrede-suvun kopiokirja. The National Archives of Finland, Helsinki. (Manuscript)
Anonymous (1653) Anno 1653 denn 12 maji. Kymenkartanon läänin renovoidut tuomiokirjat. The National Archives of Finland, Helsinki. (Manuscript)
Anonymous (1658) Anno 1653 den 21 och 22 maji. Kymenkartanon läänin renovoidut tuomiokirjat. The National Archives of Finland, Helsinki. (Manuscript)
Anonymous (1659) Denn 31 maij och den 1 junji 659. Wrede-suvun kopiokirja. The National Archives of Finland, Helsinki. (Manuscript)
Anonymous (1679) Anno 679 dhen 26 novembris. Wrede-suvun kopiokirja. The National Archives of Finland, Helsinki. (Manuscript)
Anonymous (1777) In: Harju E-S (ed) 2012: Kuninkaan kartasto Etelä-Suomesta 1776 1805. Suomal Kirjall Seur Toim 1373:1–477
Anonymous (1842) Elimäki. 3113 04 Ia. The National Archives of Finland, Helsinki. (Archival map)
Anonymous (1875) Elimäki. XII 38 Ib. Senaatin kartasto. The National Archives of Finland, Helsinki. (Archival map)
Anonymous (1980) Elimäki 3024 06. Maaperäkartta 1:20 000. Geologinen tutkimuslaitos, Maatalouden tutkimuskeskus and Maanmittaushallitus, Helsinki. (Soil map)
Anonymous (2024–2025) Last glacial period. https://en.wikipedia.org/wiki/Last_Glacial_Period
Arslanov KhA, Geĭ NA, Davydova NN, Dzhinoridze RN, Koshseckin EI, Pushsenko MYa, Rybalko AE, Spiridonov MA, Subetto DA, Khomutova VI (Арсланов ХА, Гей НА, Давыдова НН, Джинoридзе РН, Кошечкин ЕИ, Пушенко МЯ, Рыбалко АЕ, Спиридонов МА, Субетто ДА, Хомутова ВИ) (1996) Novye dannye po pozdnepleĭstochenovoi i golochenovoĭ istorii Ladozhkogo ozera (Новые данные по позднеплейстоценовой и голоценовой истории Ладожского озера). Izv Russk Geogr Obshch (Изв Русск Геогр Общ) 128(2):12–21
Arvela H, Blomqvist L, Lemmelä H, Savolainen AL, Sarkkula S (1987) Environmental gamma radiation measurements in Finland and the influence of meteorological conditions after the Chernobyl accident in 1986, vol STUK-A65. Finnish Centre for Radiation and Nuclear Safety, Helsinki, pp 1–40
Arvela H, Markkanen M, Lemmelä H (1990) Mobile survey of environmental gamma radiation and fall-out levels in Finland after the Chernobyl accident. Radiat Prot Dosim 32:177–183
Barck JF, Riddelin AJ (1835) Karta öfver Nappa bys ägor belägne i Elimä socken, Perno härad samt Nylands län. Elimäki G139, 7/6–7. The National Land Survey of Finland, Helsinki. (Archival map)
Barck JF, Unonius PL (1837) Karta öfver Korja och Eskola hemmans ägor belägne uti Philpula by Elimä socken Perno härad och Nylands län. Vilppula G139, 16/29–31. The National Archives of Finland, Helsinki. (Archival map)

Bateman MD, Boulter CH, Carr AS, Frederik CD, Peter D, Wilder M (2007) Preserving the palaeoenvironmental record in drylands: bioturbation and its significance for luminescence-derived chronologies. Sediment Geol 195:5–19. https://doi.org/10.1016/j.sedgeo.2006.07.003

Bergström EA (1703) Geometrisk delineation öfwer Umeliocki Aniala och Wärälä bÿÿars skog-segor med deras twistigheteer whilka finnes belägne uti Kymmenegårdz lähn och Elimä sochn råernas beskrifning kring alla 3 bÿars ägor. Anjala G138, 2/1–2. The National Archives of Finland, Helsinki. (Archival map)

Beta Analytic, Inc (2018) Introduction to radiocarbon determination by the accelerator mass spectrometry method. https://www.radiocarbon.cn/uploads/2018/07.pdf

Blomqvist E (1911) Kymijoki ja sen vesistö. Lisiä Suomen hydrografiaan, vol II(1). Hydrografinen toimisto, Helsinki, pp 1–235

Bonetti R, Guglielmetti A, Poli G, Sacchi E, Fink D, Hotchkis MAC, Jacobsen GE, Lawson EM, Hua Q, Smith AM, Tuniz C (1999) ^{14}C in uranium and thorium minerals: a signature of cluster radioactivity. Eur Phys J A5:235–238

Briffa KR, Bartholin TS, Eckstein D, Jones PD, Karlén W, Schweingruber FH, Zetterberg P (1990) A 1,400-year tree-ring record of summer temperatures in Fennoscandia. Nature 346:434–439

Carpelan C (1974) Hirven- ja karhunpääesineitä Skandinaviasta Uralille. Referat: Elch- und Bärenkopfgegenstände von Skandinavien bis zum Ural. Suom Museo 81:29–88

Donner J (2010) The younger Dryas age of the Salpausselkä moraines in Finland. Bull Geol Soc Finl 82:73–80. https://doi.org/10.17741/bgst/82.2.001

Eklund O (1958) Die Gefässpflanzenflora beiderseits Skiftet im Schärenarchipel Südwestfinnlands. Bidrag till kännedom af Finlands natur och folk, vol 101, pp 1–324, maps

Erä-Esko A (1958) Die Elchkopfskulptur vom Lehtojärvi in Rovaniemi. Suom Museo 65:8–18

Erichzsonn H (1638) Åhr efter Christj börds 1638 denn 15 martij. Wrede-suvun kopiokirja. The National Archives of Finland, Helsinki. (Manuscript)

Erichzsonn H (1639) Åhr efter Christj börds 1639 den 13 decembris. Wrede-suvun kopiokirja. The National Archives of Finland, Helsinki. (Manuscript)

Erichβon P (1667) [Letter]. Wrede-suvun kopiokirja. The National Archives of Finland, Helsinki. (Manuscript)

Eriksson B (1987) Siuntion Ca-konkreetion ja sitä ympäröivän saven siitepölyanalyysi. Geological Survey of Finland. https://tupa.gtk.fi/raportti/arkisto/p13_6_1_015.pdf

Eronen M (1976) A radiocarbon-dated Ancylus transgression site in South-Eastern Finland. Boreas 5:67–76

Eronen M, Haila H (1982) Shoreline displacement near Helsinki, southern Finland, during the Ancylus Lake stage. Ann Acad Sci Fenn Geol–Geogr AIII 134:111–129

Eronen M, Glückert G, Hatakka L, van der Plassche O, van der Plicht J, Rantala P (2001) Rates of Holocene isostatic uplift and relative sea-level lowering of the Baltic in SW Finland based on studies of isolation contacts. Boreas 30:17–30

Fagan B (2000) The Little Ice Age. How climate made history 1300–1850. Basic Books, New York

Frisius A (1752) Charta öfwer twistige rå linierne emillan kärande sidan Rahicka eller Nappa af Elimä sochn, och Keltis by swarande, uti Itis sochn, och Kymmene härad belägna, som är år 1752, författad af undertecknad. Elimäki G139, 7/1-2. The National Land Survey of Finland, Helsinki. (Archival map)

Frisius A (1758) Charta öfver Inginma utjord. Elimäki G139, 7/4-5. The National Land Survey of Finland, Helsinki. (Archival map)

Frisius A (1765) Charta öfwer rågången emellan Napa med Philpula af Elimä sockn å ena och Kelltis by jämte sine öfrige medintressante ifrån Ithis sokn å andra sidan. Vilppula G139, 16/1. The National Archives of Finland, Helsinki. (Archival map)

Ganander C (1789) Mythologia fennica. Christfrid Ganander and Frenckellska Boktryckeriet, Åbo. The 4th facsimile print (1984). Suomalaisen Kirjallisuuden Seura, Helsinki

Giménez-Benavides L, Escudero A, Pérez-García F (2005) Seed germination of high mountain Mediterranean species: altitudinal, intrapopulation and interannual variability. Ecol Res 20:433–444. https://doi.org/10.1007/sl1284-005-0059-4

Goodship A, Alexanderson H, Anjar J (2025) Luminescence analysis of a postglacial landscape in northeastern Norway and northern Finland. Boreas 55(1):164–181. https://doi.org/10.1111/bor.7001570015

GTK (1992–2001) Aeroradiometric thorium map of Finland. Aeroradiometric uranium map of Finland. Spatial data products. The Geological Survey of Finland, Espoo

Halme E (1969) Kyminkartanon läänin kalastus 1500- ja 1600-luvulla. MSc thesis, University of Helsinki

Hartl D (1966) Pseudolysimachion spicatum (L) Opiz. In: Hartl D, Wagenitz G (eds) Hegi illustrierte Flora von Mitteleuropa VI, vol 1(2). Carl Hanser Verlag, Munich, pp 153–155

Hartmann HD (1966) Stickstoffdüngungsversuche zu Schnittlauch. Gartenbauwissenschaft 31:51–63

Hausen G (1920) Nylands ortnamn. Skrift utgiv Svenska Litteratursällsk Finl 152:I–XIX, 1–318

Hemmilä R (2017) Pommi pohjassa. Koala-Kustannus, Helsinki

Hemmilä R (2018) Keltist Mylsään. R Hemmilä, Koria

Hinrichson H (1635) Å. 1635 den 14 maij. Wrede-suvun kopiokirja. The National Archives of Finland, Helsinki. (Manuscript)

Hipping AJ (1817) Beskrifning öfwer Perno socken i Finland. MC Iversen, St. Petersburg

Hughes ALC, Gyllencreutz R, Lohne ØS, Mangerud J, Svendsen JI (2016) The last Eurasian ice sheet—a chronological database and time-slice reconstruction, DATED-1. Boreas 45:1–45, e1–e22. https://onlinelibrary.wiley.com/doi/epdf/10.1111/bor.12142

Hyyppä E (1937) Post-glacial changes of shore-line in South Finland. Bull Commis Géol Finl 120:1–225, 2 appendices

Hyyppä E (1942) Beiträge zur Kenntnis der Ladoga- und Ancylustransgression. Compt Rend Soc Géol Finl 15:139–178. https://www.geologinenseura.fi/sites/geologinenseura.fi/files/bt_128.pdf

Hyyppä E (1950) Helsingin ympäristön maaperäkartan selitys. Geologinen tutkimuslaitos, Helsinki. (Commentaries of the Helsinki soil map)

Hyyppä E (1958) Geologische Datierung der Elchkopfskulptur vom Lehtojärvi bei Rovaniemi. Suomen Museo 65:19–25

Hyyppä E, Hoffrén V, Isola A (1962) Geological Survey of Finland radiocarbon measurements. I. Radiocarbon 4:81–83

Hyyppä E, Toivonen AVP, Isola A (1964) Geological Survey of Finland radiocarbon measurements. III. Radiocarbon 6:110–111

Hyyppä E, Heikkinen A, Toivonen V (1969) Geological Survey of Finland radiocarbon measurements. IV. Radiocarbon 11:183–187

Ioansson B (1380) Vid Åby kyrka 26 februari 1380. In: Hausen R (ed) (1910) Finlands medeltidsurkunder, vol 1, p 377. https://df.narc.fi/Images/Editions/fmu/Hausen_FMU_1.pdf

Israellsonn [Israelβon] A (1640) Anno Christj 1640 den 31 augustj. Wrede-suvun kopiokirja. The National Archives of Finland, Helsinki. (Manuscripts)

Itkonen E, Joki AJ, Peltola R (1975) Suomen kielen etymologinen sanakirja. Part 5. Lexica Societatis Fenno-Ugricae XII(5):1257–1676

Iversen J (1954) The late-glacial flora of Denmark and its relation to climate and soil. Geol Surv Denm (Ser II) 80:87–119

Jenkinson A (1579) Russie, moscoviae et tartarie description. In: Theatrum Orbis Terrarum. Abraham Ortelius, Amsterdam. (Facsimile map)

Jull AJT, Barker DL, Donahue DJ (1987) On the ^{14}C content in radioactive ores. Chem Geol (Isotope Geosci Sect) 66:35–40

Kääriäinen E (1963) Land uplift in Finland computed by the aid of precise levellings. Fennia 89(1):15–19, map

Kaipainen M (2022) Kuivatusta järvestä tuli viljelijöiden murheenkryyni – pellot muuttuvat aina takaisin järveksi. https://yle.fi/a/3-12427547

Kakkuri J (1987) Character of the Fennoscandian land uplift in the 20th century. Geol Surv Finl Spec Pap 2:15–20. https://tupa.gtk.fi/julkaisu/specialpaper/sp_002.pdf

Kare E (1929) Valamo. Maanmittaushallitus, Helsinki. (Topographical map 1:20 000). https://www.Karjalankartat.fi

Kitagawa H, van der Plicht J (1998) Atmospheric radiocarbon calibration to 45,000 yr. B.P.: late glacial fluctuations and cosmogenic isotope production. Science 279:1187–1190

Kivikäs P (1995) Kalliomaalaukset muinainen kuva-arkisto. Summary: Paintings on rock—an ancient picture archive. Atena Kustannus Oy, Jyväskylä, pp 1–324, 334–336

Kokkonen P (1926) Kymijoen perkauksesta johtuva maanarvon nousu. Referat: Die Steigerung des Bodenwertes infolge der Reinigung des Kymijokiflusses. Suomen Maataloustiet Seur Julk 11:1–46

Kokkonen H (1927) Elimäen pitäjän historia. Part I. Elimäen kunta, Elimäki

Kokorev AV (2003) Water balance of Lake Ilmen. In: Simola H, Terzhevik AYu, Viljanen M, Holopainen IJ (eds) Proceeding of the fourth international Lake Ladoga symposium 2002, vol 138. University of Joensuu, Karelian Institute, pp 370–375

Kujala V, Ulvinen A (1964) Floristische Untersuchungen in Ost-Kymenlaakso in Südfinnland. Ann Bot Soc Zool Bot Fenn Vanamo 35(2):1–215

Kunttu P (2019) Vuorimunkilla vahva kanta Taalintehtaalla – suojelu ja hoito silti ajankohtaista. Abstract: The population of Jasione montana is strong in Dalsbruk (SW Finland) but habitat protection is needed. Lutukka 35:78–82. www.luomus.fi/lutukka

Kurck J (1646) Den konungzlige rätts i Finland äntelige sentens och dom uthi den langwarlige twist, emellan dee Elimä frälβe bönder. Wrede-suvun kopiokirja. The National Archives of Finland, Helsinki. (Manuscript)

Kuusisto E (1982) Euroopan suurimman järven hydrologiaa. Vesitalous 23(1):10–12

Kvasov DD, Nazarenko VA (Квасов ДД, Назаренко ВА) (1970) O datirovke maksimuma Ladozhskoĭ transgressii (О датировке максимума Ладожской трансгрессии). In: Kabaĭlene MV (ed) Seengeschichte. Berichte des für die Hauptfragen des Süsswasserseen gewidmeten Allunionsymposiums. Vilnius, 25-29 Mai 1970. Ežerai-70. (История озер. Труды Всесоюзного симпозиума), vol 2, pp 332–341

Lampinen R, Lahti T (2019) Kasviatlas 2018. Helsingin Yliopisto, Luonnontieteellinen keskusmuseo, Helsinki. http://koivu.luomus.fi/kasviatlas. (Plant distribution maps and database)

Larβon J (1650) Anno 1650 denn 16 januarij. Wrede-suvun kopiokirja. The National Archives of Finland, Helsinki. (Manuscript)

Larβonn J (1653) Anno 1653 den 18 och 19 novemb. Wrede-suvun kopiokirja. The National Archives of Finland, Helsinki. (Manuscript)

Lewong J (1772) Vilppula och Rahikkala rågångskarta med beskrivning. Vilppula G139, 16/3-4. The National Archives of Finland, Helsinki. (Archival map)

Lewong J (1773) Charta öfwer Hapasuo ängs rödningar å Philpola bys ägor i Elime sochn Kymmene härad och Kymmenegårds län. Elimäki G139, 16/5. The National Archives of Finland, Helsinki. (Archival map)

Lisitsyna GN (Лисицына ГН) (1961) Voprosy paleogeografii neolita raĭonov Severo-zapada Evropeĭskoĭ chasti SSSR (Вопросы палеогеографии неолита районов Северо-запада Европейской части СССР) Mater issl arheol SSSR (Матер иссл археол СССР) 87:535–578

Liiva A, Ekman I, Rinne T (1978) Tartu radiocarbon dates. VII. Radiocarbon 20:441–448

Lopmeri JA (1947) Vanhan Vehmaan kihlakunnan pitäjien ja kylien nimet. Suomal Kirjall Seur Toim 220(2):1–210

Ludikova AV, Subetto DA, Kuznetsov DD, Orlov AV, Shatalova AE (2024) New diatom and sedimentary data confirm the existence of the northern paleo-outlet from Lake Ladoga to the Baltic Sea. Quaternary 7(3):31. https://doi.org/10.3390/quat7030031

Luukkonen IM (1906–1907) Pyykeillä n:o 1–5 rajoitettu alue on pakkoluovutettu vuonna 1906. Vilppula G139, 16/29–31. The National Archives of Finland, Helsinki. (Archival map)

Luukkonen IM, Luostarinen H (1913–1915) 3-osainen kartta Vilppulan kylän kaikista tiluksista Elimäen pitäjässä Pernajan kihlakunnassa Uudenmaan lääniä. Vilppula G139, 16/40–106. The National Archives of Finland, Helsinki. (Archival map)

MapSite (2022–2024) The National Land Survey of Finland, Helsinki. maanmittauslaitos.fi/en/e-services/mapsite

Meurman I (1908–1911) Uudistuskartta Moision ja Rahikkalan kylien kaikista maista Elimäki G139, 10/5-6. The National Archives of Finland, Helsinki. (Archival map)

Miettinen T (2000) Kymenlaakson kalliomaalaukset. Abstract: Rock paintings in Kymenlaakso. Kymenlaakson maakuntamuseon julkaisuja 27:1–163

Misra A, Tyler G (1999) Influence of soil moisture on soil solution chemistry and concentrations of minerals in the calcicoles Phleum phleoides and Veronica spicata grown on a limestone soil. Ann Bot 84:401–410. http://www.idealibrary.com

Moberg A, Sonechkin DM, Holmgren K, Datsenko NM, Karlén W (2005) Highly variable Northern Hemisphere temperatures from low- and high-resolution proxy data. Nature 433:613–617, 1014. https://doi.org/10.1038/nature03265

Moore AMT, Kennett JP, Napier WM, LeCompte MA, Moore CR, West A (2023) Abu Hureyra, Syria, Part 3: comet airbursts triggered major climate change 12,800 years ago that initiated the transition to agriculture. Airbursts Cratering Impacts 1:1–24. https://doi.org/10.14293/ACI.2023.0004

Munselius C (1666) Anno 1666 den 14, 15 maij. Wrede-suvun kopiokirja. The National Archives of Finland, Helsinki. (Manuscript)

Nimisampo (2024) Names Archive. Institute for the Languages of Finland, Helsinki. https://nimisampo.fi

Oinonen M, Alenius T, Arppe L, Bocherens H, Etu-Sihvola H, Helama S, Huhtamaa H, Lahtinen M, Mannermaa K, Onkamo P, Palo J, Sajantila A, Salo K, Sundell T, Vanhanen S, Wessman A (2020) Buried in water, burdened by nature—resilience carried the Iron Age people through fimbulvinter. PLoS One 15(4) e0231787. https://doi.org/10.1371/journal.pone.0231787

Ojala AEK, Palmu J-P, Åberg A, Åberg S, Virkki H (2013) Development of an ancient shoreline database to reconstruct the Litorina Sea maximum extension and the highest shoreline of the Baltic Sea basin in Finland. Bull Geol Soc Finl 85:127–144. https://doi.org/10.17741/bgsf/85.2.002

Okko V (1964) Maaperä. In: Rankama K (ed) Suomen geologia. Kirjayhtymä, Helsinki, pp 239–332

Oksanen E-L (1981) Anjalan historia. Anjalan kunta, Anjalankoski, pp 13–14, 29–691

Oksanen E-L (1985) Elimäen historia. Elimäen kunta, Elimäki

Ottelin A (1793) Storskiftes delnings beskrifning öfver Moisio sätesgårds, med derunder lydande rå och rörs hemmans torp ängar och pares, samt Rahickala och Jåkela frälsehemmanens åkrar, hafverländer. The Municipal Archive, Elimäki. (Manuscript)

Ottelin A, Kihlström O (1792–1810) Karta öfwer Moisio och Rahickala byars ägor uti Elimä socken af måtte till storskifte 1792. Elimäki C 3.19. The National Land Survey of Finland, Kouvola. (Archival map)

Pakkala EW (1902) Luettelo Pernajan kihlakunnan suomalaisten pitäjien muinaismuistoista. Suomen Muinaismuistoyhdistyksen pöytäkirjat. 10/X 1902. www.kyppi.fi/to.aspx?id=129.146408. (Manuscript)

Panova E, Vlasov D, Luodes H (2014) Evaluation of the durability of granite in architectural monuments. Geol Surv Finl Rep Invest 214:1–79, 3 appendices. tupa.gtk.fi/julkaisu/tutkimusraportti/tr_214.pdf

Paulaharju S (1932) Härmän aukeilta. Werner Söderström, Porvoo. https://doria.fi/handle/1932_f470726.pdf

Pedherson L (1568) Anno Dni 1568 then 15 dagen julij. Wrede-suvun kopiokirja. The National Archives of Finland, Helsinki. (Manuscript)

Pesonen P (2003) Esihistoria. In: Aalto S (ed) Pornaisten historia 1. Pornaisten kunta, Pornainen, pp 23–55, 299–301 and 309–315

Pfäffli EA, Hellman I (1918) 14-osainen kartta Niinimäen tilojen RN№ 1^2–4^{62} kaikista tiluksista Iitin kunnassa Uudenmaan lääniä. Iitti G140, 10/78-169. The National Archives of Finland, Helsinki. (Archival map)

Pohjakallio O (1941) Sinimailasen ja rantamaitteen viljelymahdollisuuksista Suomessa Referat: Über die Anbaumöglichkeiten der Luzerne und des gemeinen Hornklees in Finnland. Valt maatalouskoetoim julk. Die staatl landwirtschaft Versuchstätigk Veröffentl, vol 110, pp 1–23. valmtkt_julk110.pdf

Pöllänen R, Valkama I, Toivonen H (1997) Transport of radioactive particles from the Chernobyl accident. Atmos Environ 31:3575–3590

Punkari M, Boulton GS (1995) Skandinavian mannerjäätikön itäosan dynamiikka ja reuna-asemat nuoremman dryaksen aikana. Summary: The dynamics and original positions of the Scandinavian ice sheet in the eastern part during the Younger Dryas stadial. Terra 107:3–14

Punttila M (1960) Karjusuo and Mannari, Elimäki. Names Archive. Institute for the Languages of Finland, Helsinki. (Archival record)

Reinholm HA (1839) H. A. Reinholmin arkisto. Suomen Muinaismuistoyhdistys. Archives of the National Board of Antiquities, Helsinki. (Manuscripts)

Saarnisto M (1970) The late Weichselian and Flandrian history of the Saimaa Lake complex. Soc Sci Fenn Comment Phys Math 37:1–107, appendices

Saarnisto M, Grönlund T (1996) Shoreline displacement of Lake Ladoga—new data from Kilpolansaari. Hydrobiologia 322:205–215

Seppä H, Birks HJB (2001) July mean temperature and annual precipitation trends during the Holocene in the Fennoscandian tree-line area: pollen based climate reconstruction. Holocene 11:527–539

Seppovaara O (1988) Kymijoki. Kouvolan Kirjapaino Oy, Kuusankoski

Sigfwersson P (1544) Anno Dni 1544 then 25 Septemb. Wrede-suvun kopiokirja. The National Archives of Finland, Helsinki. (Manuscript)

Sirén A (1963) Recent land uplift in Finland computed from lake water level records. Fennia 89(1):11–13

Starck CA, Bonsdorf GEE (1828–1831) Charta öfver Philpula byss ägor belägne uti Kymmenegårdslän. Elimäki C.6.41. The Geological Survey of Finland, Helsinki. (Archival map)

Suojala T (2003) Yield potential of chive: effects of cultivar, plastic mulch and fertilization. Agric Food Sci Finl 12:95–105. https://doi.org/10.23986/afsi.5745

Teitti J (1555–1556) Jakob Teitts klagomålsregister emot adeln i Finland. Microfilms 215j and 215k. The National Archives of Finland, Helsinki. (Manuscript)

Tikkanen M, Oksanen J (2002) Late Weichselian and Holocene shore displacement history of the Baltic Sea in Finland. Fennia 180:9–20. CD. https://fennia.journal.fi/article/download/3760/3551

Tiljander M, Saarnisto M, Ojala AE, Saarinen T (2003) A 3000-year palaeoenvironmental record from annually laminated sediment of Lake Korttajärvi, Central Finland. Boreas 6:566–577

Toivonen YH, Itkonen E, Joki AJ (1958) Suomen kielen etymologinen sanakirja. Part 2. Lexica Societatis Fenno-Ugricae XII(2):205–480. The 5th unchanged reprint (1994)

Toivonen H, Servomaa K, Rytömaa T (1988) Aerosols from Chernobyl: particle characteristics and health implications. In: Philipsborn Hv, Steinhäusler F (eds) Hot particles from the Chernobyl fallout. Schriftenreihe des Bergbau- und Industriemuseums Ostbayern Theuern 16:97–105

Törn JE (1753) År 1753 den 7 etc. januarii, blef lagtima härads winter ting hållit med allmogen och them menige man af Nyby och Itis sochnars tingslag, uti Willächtis by. Elimäki G139, 7/3. The National Archives of Finland, Helsinki. (Manuscript)

Tynni R, Hyyppä J, Valovirta V (1976) Lapinjärven kartta-alueen maaperä. Summary: quaternary deposits in the Lapinjärvi map-sheet area. Suomen geologinen kartta 1:100 000. Maaperäkartan selitykset, 3022 Lapinjärvi. Geologinen tutkimuslaitos, Espoo

Ukkonen P, Mannermaa K (2017) Museoviraston julkaisuja, vol 8, pp 1–240

Ukkonen P, Lunkka JP, Jungner H, Donner J (1999) New radiocarbon dates from Finnish mammoths indicating large ice-free areas in Fennoscandia during the Middle Weichselian. J Quat

Sci 14:711–714. https://doi.org/10.1002/(SICI)1099-1417(199912)14:7<711::AIDJQS506>3.0.CO;2-E

Ukkonen P, Arppe L, Houmark-Nielse M, Kjær KH, Karhu JA (2007) MIS 3 mammoth remains from Sweden—implications for faunal history, palaeoclimate and glaciation chronology. Quat Sci Rev 26:3081–3098

Vainio A (1918) (Names concerning the Värälä village in Elimäki) Names Archive. Institute for the Languages of Finland, Helsinki. (Archival records)

Väliranta M, Salonen JS, Heikkilä M, Amon L, Helmens K, Klimaschewski A, Kuhry P, Kultti S, Poska A, Shala S, Veski S, Birks HH (2015) Plant macrofossil evidence for an early onset of the Holocene summer thermal maximum in northernmost Europe. Nat Commun 6:6809. https://doi.org/10.1038/ncomms7809

van Geel B, Buurman J, Waterbolk HT (1996) Archaeological and palaeoecological indications of an abrupt climate change in The Netherlands, and evidence for climatological teleconnections around 2650 BP. J Quat Sci 11:451–460

Vasmer M (1953) Russisches etymologisches Wörterbuch. A–K, vol 1: I–XLVII and 1–712. Carl Winter Universitätsverlag, Heidelberg

Wanner H, Pfister C, Neukom R (2022) The variable European Little Ice Age. Quat Sci Rev 287:107531. https://doi.org/10.1016/j.quascirev.2022.107531

Westermark P (1773) Charta beskrifning öwer Anjala och Ummeljoki råskillnader uti Elimä sokn. Elimäki G138, 2/3-4. The National Archives of Finland, Helsinki. (Archival map)

Westermark P (1774–1776) Charta öfver Anjala, Ummeljoki och Ahwis byars ägor uti Elimä sokn, Kymmene härad samt Kymmenegårds höfdingedöme. Elimäki G138, 1/2-11. The National Archives of Finland, Helsinki. (Archival map)

Westermark P, Barck JF (1774–1848) Karta öfver Ahwis bys ägor uti Elimä soken, och Nylands län. Elimäki G138, 1/12-14. The National Archives of Finland, Helsinki. (Archival map)

Znamenskaya OM, Sokolova VB, Khomutova VI (Знаменская ОМ, Соколова ВБ, Хомутова ВИ (1970) Sravnitel'nyĭ analiz paleogeograficheskih usloviĭ razvitiya yuzhnyx i zapadnyh beregov Ladozhskogo ozera (Сравнительный анализ палеогеографических условий развития южных и западных берегов Ладожского озера) In: Kabaĭlene MV (ed) Seengeschichte. Berichte des für die Hauptfragen des Süsswasserseen gewidmeten Allunionsymposiums. Vilnius, 25–29 Mai 1970. Ežerai-70. (История озер. Труды Всесоюзного симпозиума), vol 2, pp 319–331

Chapter 5
Conclusions

Abstract Samples of the littoral inland plant stands have survived on the shores since the Yoldia Sea, altitudes for older seashore levels hardly occur in the region studied. The Kymi Dale forms an anomaly of land upheaval compared with its eastern and western flanks. Recorded oral Finnish traditions have borne in mind events which can now be physically dated. The knowledge that the Elimäki territory has been 'part of the great bottom of the Kymi' might originate from the outburst of the river-to-be through the First Salpausselkä ridge, now dated to have occurred 7.61 ± 0.46 ka ago. The present channel of the Kymi did not exist then, and flood prevailed. The floods were ancient geohazards. The tradition that the sea extended to the Anjalankoski (syn. Ankkapurha) rapids might bear a memory from the X sea stage, now dated to have begun about 7.4 cal ka BP. The transgression of the Ladoga remembered in the local oral Finnish tradition on the Mantsinsaari Island in the 1800's, keenly knew the transgression to be a consequence of the lack of an outlet of the Ladoga which state ended about 4.1 cal ka BP. The longevity of the spoken Finnish language in the region is older than the linguists have thought.

Keywords Age of the finnish language · Age of sand eskers · Ice blockages in the kymi river · Inland littoral plant species · Land upheaval differences · Mantsinsaari Island · Oral tradition · Supercooling · Ummeljoki

The examined region has been formed by the interactions of eustatic level changes in the Baltic Sea and, additionally, by isostatic sink and upheaval of the land during and after the Weichselian Ice Age. Uneven sink and upheaval rates resulted in the Kymi Dale (Kymenlaakso) and in the inclining of the whole Finland-to-be towards the southeast. Upon these gradual changes, more sudden irregularities of the macroclimate have reduced the incident solar radiation to the globe and caused variations in precipitation. The reasons for the macroclimate changes are under debate.

H. Ahokas et al., *Paleoshorelines and Their Relict Littoral Plants*, SpringerBriefs in Earth Sciences, https://doi.org/10.1007/978-3-032-17723-0_5

The glaciofluvial sand samples from the Välikangas esker in S Kouvola were luminescence dated to the minimum ages of 63 ± 3 and 56 ± 2.7 ka, which represent the melting during the Middle Weichselian, leading to the ice-free stadial in central Finland. Stormy weather between 11.5 ± 0.4 and 9.5 ± 0.4 ka ago caused sand redeposition on the flanks of the glaciofluvial eskers. It is in these redeposited sands that preserved evidence of paleoshorelines occurs in the form of raised beaches. The raised beaches in focus here indicate sea levels ranging from 35 to 43 m asl. The appearance of the beaches was dated with radiocarbon to between 7.4 and 3.8 cal ka BP.

Four relic littoral plant species sporadically occur on the ancient shores. The species are chive (*Allium schoenoprasum*), sheep's-bit (*Jasione montana*), trefoil (*Lotus corniculatus*), and spiked speedwell (*Veronica spicata*). Because the highest altitudes of their occurrences correspond to the Yoldia shore levels, they probably all occurred here as pioneer species on the ancient shores of the Yoldia Sea 11.2–10.2 cal ka BP, where open, non-shaded conditions were favourable to them. Reed (*Phragmites australis*), a seaside plant species, has survived in many places in inland sites as a relic of the past shore. Its survival is probable in an environment of pH 5.9 or more and rich with basic cations. Here, two field populations of reed on the Saarela farm (in Rahikkala village) were dated with radiocarbon of the nearby raised beach of the undefined X sea to have occurred since 6.88–6.68 cal ka BP.

An outburst of flood southwards of the Kymi river through the First Salpausselkä ridge at the Ropakonmäki esker formed the Koria sand plains <2 km southwards at 7.61 ± 0.46 ka ago. This was evidently preceded by a period of rainy climate. Radiocarbon dating also suggests major streaming in the Koria sands still between 3.65 and 2.0 cal ka BP. More recent floods caused by ice blocking the river narrows are documented in oral traditions, documentary records, dating, and some toponyms, e.g., Ummikoski ('Blocked rapids'; 'Vmikoski' in 1593; Hausen 1920), a former name of the Ummeljoki ('Blocked river') village. Supercooled water evidently caused sudden and severe blockages in rivers, followed by floods. Some floods are thought to have been consequences of cold or rainy epochs between 2.3 and 1.9 cal ka BP, at *fimbulvinteren c.* 1.41 cal ka BP, and during the Little Ice Age at *c.* 350 cal a BP. The current course of the Kymi River, which flows downwards from the Second Salpausselkä ridge, aligns with the maxima of the anomalously higher land upheaval rates in its valley, which were levelled between 1935 and 1955.

Especially fishing, sealing, and beaver hunting, as well as opportunities for boat transportation, bound the ancient people to waterways. The danger of uncontrollable floods had to be considered when selecting dwelling sites. In the region examined here, elevations safe from dangerous floods seem to have often been given names meaning 'island,' though they were surrounded by firm land, or received names referring to various kinds of elevations. Reinholm (1839–) even wrote down the oral tradition in this region, that the ancient people had lived on hills during the flooding era.

The linguistic tradition carried over from major geological events demonstrates the lineage of a common, understandable language for millennia in this region, where Finnish happens to be the only historically known ancient language. Four cases for the physical dating of the spoken Finnish are listed here. (1) Oral Finnish traditions might have known the outburst of the Kymi, describing the Elimäki territory to have been 'part of the great bottom of the Kymi' (Pakkala 1902). Such spreading of the water was possibly initiated by the outburst that occurred 7.61 ± 0.46 ka ago, forming the Koria sand plains, as shown in this book. (2) Reinholm (1839–) wrote down the oral tradition that a seashore extended to the current inland, up to Ankkapurha (Fig. 1.1) in the past. If it meant the Litorina Sea, the shore may have been nearby Ankkapurha (Hyyppä 1937), 6.3 ka ago or earlier (Miettinen 2002). If it were the X sea examined in this book, such a situation may have occurred as late as 4.06–3.83 cal ka BP, based on radiocarbon dating from Pakinkangas (Table 3.5). (3) The old Finnish folklore has about 300 archival records that describe or mention the flying of a conical object from the northeast. The object could have ignited a fire on land, and eventually, a part of it impacted the sea. These verses probably describe a meteorite that split into smaller impactors in the sky while approaching from the northeast towards the southern coast of Finland, circa. 6.4–6.2 ka ago (Ahokas 2023). (4) Ailio (1897) recorded from the Mantsinsaari island on the Ladoga the local Finnish oral tradition, which knew the Ladoga to have been blocked without an outlet in the past. This oral tradition from Mantsinsaari described the millennia-old, remembered situation. Physically dated, the blocked situation ended with a temporary outlet by *c.* 4.1 cal ka BP (Ludikova et al. 2024), and thereafter with the Neva outlet river (Saarnisto and Grönlund 1996; Ludikova et al. 2024).

References

Ahokas H (2023) Thirty-three previously unknown meteoritic craters of diameter from 5 to 100 m in western Kouvola, Finland from a swarm of impactors in the Holocene. Kave, Helsinki. https://www.impact-structures.com/finland

Ailio J (1897) Über Strandbildungen des Litorinameeres auf der Insel Mantsinsaari. Fennia 14(2):1–43, map

Hausen G (1920) Nylands ortnamn. Skrift utgiv Svenska Litteratursällsk Finl 152:I–XIX, 1–318

Hyyppä E (1937) Post-glacial changes of shore-line in South Finland. Bull Commis Géol Finl 120:1–225, 2 appendices

Ludikova AV, Subetto DA, Kuznetsov DD, Orlov AV, Shatalova AE (2024) New diatom and sedimentary data confirm the existence of the northern paleo-outlet from Lake Ladoga to the Baltic Sea. Quaternary 7(3):31. https://doi.org/10.3390/quat7030031

Miettinen A (2002) Relative sea level changes in the eastern part of the Gulf of Finland during the last 8000 years. Ann Acad Sci Fenn Geol–Geogr 162:1–100

Pakkala EW (1902) Luettelo Pernajan kihlakunnan suomalaisten pitäjien muinaismuistoista. Suomen Muinaismuistoyhdistyksen pöytäkirjat. 10/X 1902. www.kyppi.fi/to.aspx?id=129.146408. (Manuscript)

Reinholm HA (1839) H. A. Reinholmin arkisto. Suomen Muinaismuistoyhdistys. Archives of the National Board of Antiquities, Helsinki. (Manuscripts)

Saarnisto M, Grönlund T (1996) Shoreline displacement of Lake Ladoga—new data from Kilpolansaari. Hydrobiologia 322:205–215

Chapter 6
Supplementary Information

Abstract Details of the luminescence dating of the samples are presented.

Keywords Ahkoja-L-1 Shfd16169 · Kaurisselkä-L NCL-7320041 · Pakinkangas-L NCL-7320042 · Raavuori-L NCL-7320043 · Välikangas-1-L NCL-7320044 · Välikangas-2-L NCL-7320045

6.1 Sheffield Luminescence Dating

Mark D. Bateman

See Figs. 6.1, 6.2, 6.3 and Table 6.1.

6.2 NCL Luminescence Dating

Elizabeth L. Chamberlain

See Tables 6.2 and 6.3 and Figs. 6.4 and 6.5.

Figures 6.6, 6.7, 6.8, 6.9, and 6.10 Equivalent dose distributions of pIRIR-225 signals obtained from single grains are presented below as radial plots (Galbraith 1990) for each sample:

H. Ahokas et al., *Paleoshorelines and Their Relict Littoral Plants*, SpringerBriefs in Earth Sciences, https://doi.org/10.1007/978-3-032-17723-0_6

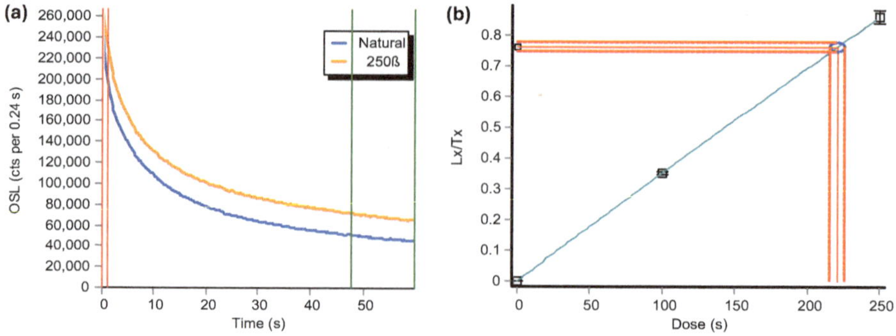

Fig. 6.1 (**a**, **b**) Example of single aliquot OSL data for sample Shfd16169. (**a**) OSL decay of naturally acquired and Beta dose regenerated signal. High initial signal above background which decays moderately away to background may have some unstable medium component problems but these did not impact on the dose recovery test which was able to cover well a known dose. (**b**) SAR growth curve showing good growth was laboratory dose and interpolation of the naturally acquired OSL

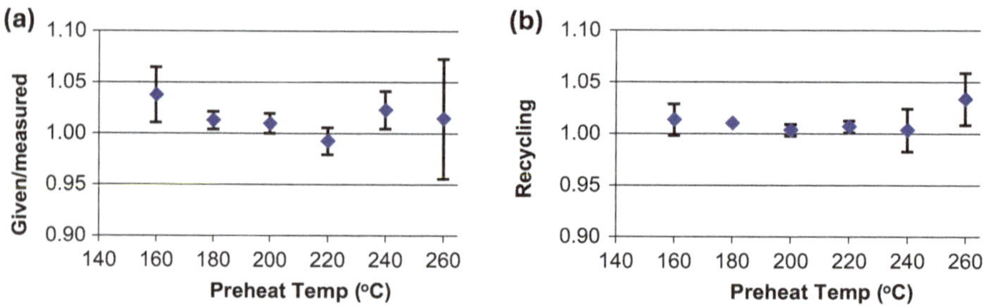

Fig. 6.2 (**a**, **b**) Results of different preheated temperatures in recovering a *c.* 20 Gy beta radiation dose. (**a**) Given to recovered dose ratio at different preheat temperatures. (**b**) Recycling ratio (ratio between the first and last dose point) at different preheat temperatures. Data points in both plots are the averages of three measurements performed for each preheat temperature

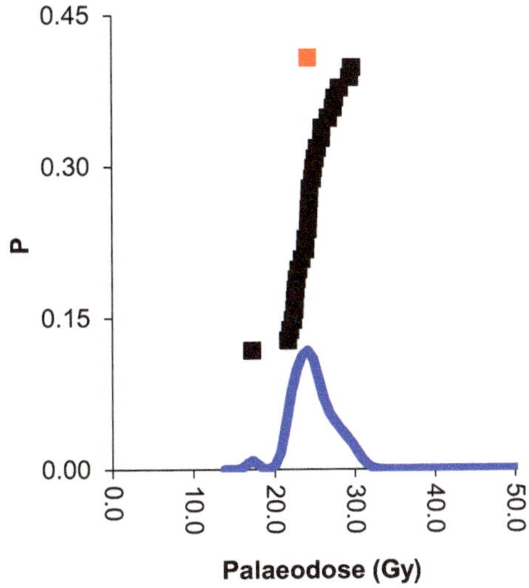

Fig. 6.3 A combined probability density function of De values from single aliquot measurements for sample Shfd16169 showing the low degree of inter-aliquot variability. Also plotted are individual aliquot De values (black) and the unweighted mean De (red)

Table 6.1 Summary of results. Dosimetry related data

Lab code	U (ppm)	Th (ppm)	K (%)	D_{cosmic} (μGy/a)	Moisture (%)	Dose rate[a] (μGy/a)
Shfd16169	1.53	6.1	3.0	198 ± 10	19	3196 ± 187

[a] Total Dose is attenuated for grain size, density and moisture

Table 6.2 Measurement sequence applied at the Netherlands Centre for Luminescence dating (NCL) to obtain the feldspar single grain pIRIR-225 signal, based on prior studies (Brill et al. 2018; Reimann et al. 2017)

Step	Action	Measured
1	Beta dose (or natural dose)	
2	60 s preheat to 250 °C	
3	2 s single-grain IR laser stimulation at 50 °C	
4	2 s single-grain IR laser stimulation at 225 °C	L_n, L_i
5	300 s IR stimulation at 225 °C	
6	Beta test dose	
7	60 s cut heat to 250 °C	
8	2 s single-grain IR laser stimulation at 50 °C	
9	2 s single-grain IR laser stimulation at 225 °C	T_n, T_i
10	300 s IR stimulation at 225 °C	
11	Repeat step 1–10 for a range of doses (incl. zero and repeat dose)	

Table 6.3 Detailed luminescence dating results for samples measured at NCL. Details of paleodose and dose rate analyses for luminescence samples measured at NCL including the number of grains accepted, number of grains contributing to the paleodose, cosmogenic dose rate, radionuclide activities, measured and used water contents in percent dry weight, and measured organic content in percent dry weight. All data are for the 212–250 μm feldspar sand fraction

Sample code	Grains accepted n	Grains age model n	Cosmogenic dose (Gy/ka)	U (Bq/kg)	Th (Bq/kg)	^{40}K (Bq/kg)	Water content measured (%)	Water content used (%)	Organic content (%)
NCL-7320041	246	221	0.19 ± 0.01	31.6 ± 0.5	32.7 ± 1.0	1152 ± 25	7.1	7.1 ± 1.8	0.9 ± 0.09
NCL-7320042	193	142	0.20 ± 0.01	14.8 ± 0.3	12.7 ± 0.7	1147 ± 25	8.0	8.0 ± 2.0	0.2 ± 0.02
NCL-7320043	206	166	0.21 ± 0.01	43.6 ± 0.7	52.7 ± 1.4	1469 ± 31	2.1	5.0 ± 1.3	0.3 ± 0.03
NCL-7320044	37	35	0.17 ± 0.01	40.6 ± 0.7	51.2 ± 1.3	1074 ± 23	5.5	5.5 ± 1.4	0.2 ± 0.02
NCL-7320045	60	51	0.18 ± 0.01	37.0 ± 0.6	47.4 ± 1.3	1011 ± 22	2.8	5.0 ± 1.3	0.2 ± 0.02

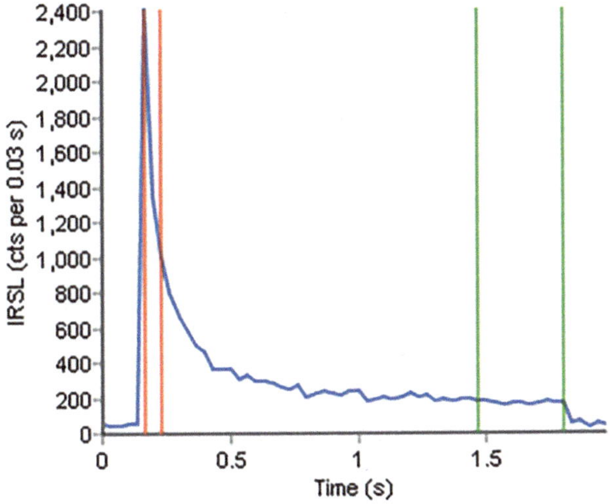

Fig. 6.4 Example of a typical pIRIR-225 single-grain feldspar decay curve for this project, obtained from sample NCL-7320043

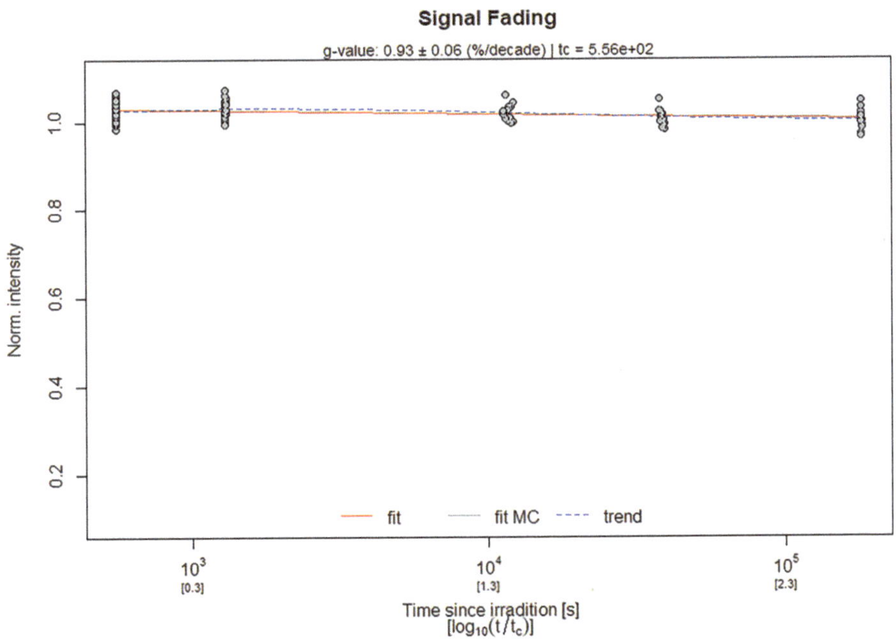

Fig. 6.5 Feldspar pIRIR-225 measured on small-diameter multi-grain aliquots at NCL showed negligible to no fading, indicating this signal is sufficiently stable for dating the samples

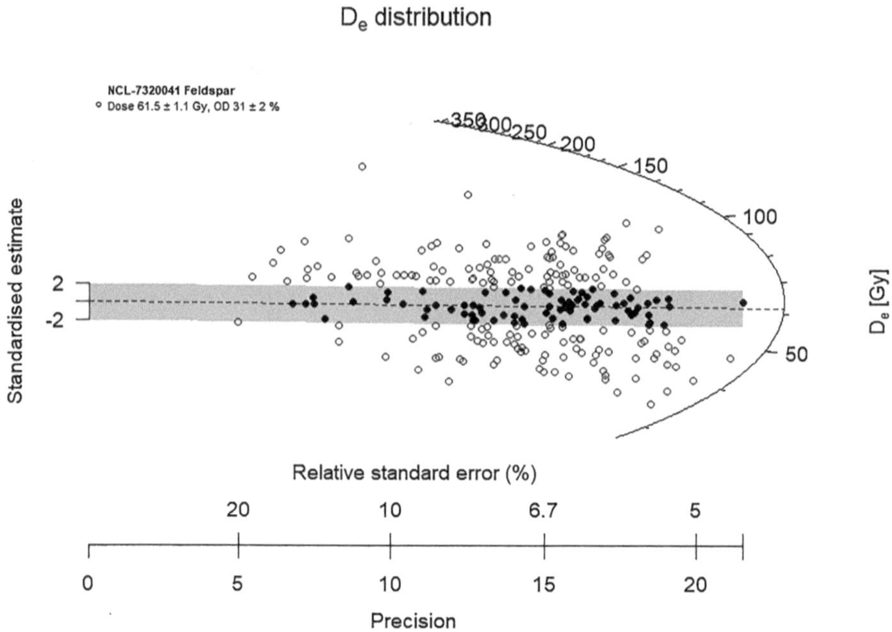

Fig. 6.6 Feldspar pIRIR-225 single-grain dose distribution for sample NCL-7320041

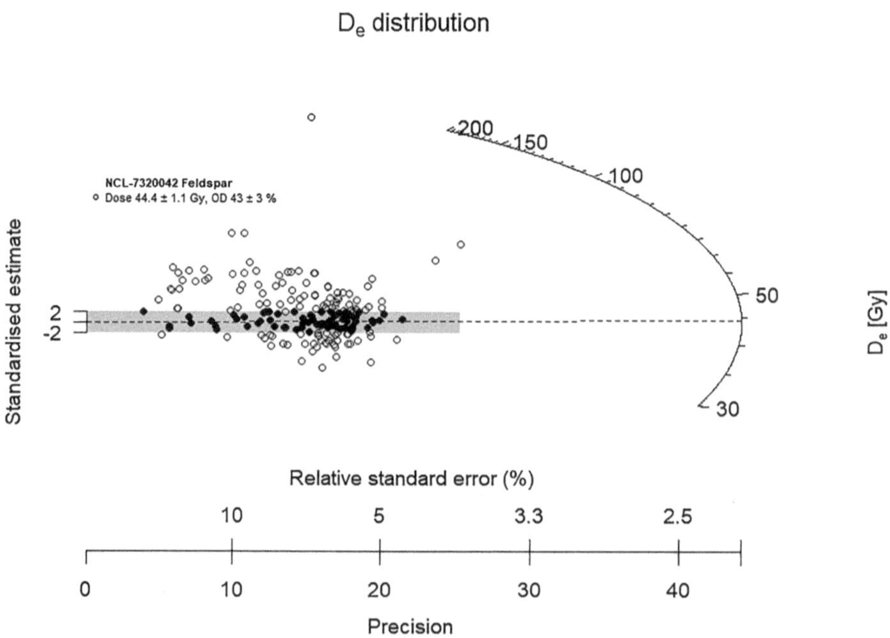

Fig. 6.7 Feldspar pIRIR-225 single-grain dose distribution for sample NCL-7320042

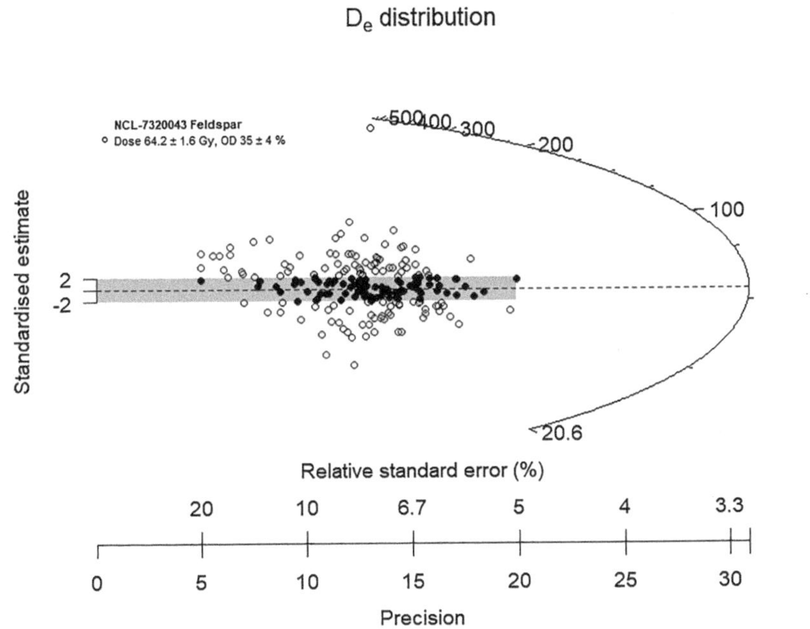

Fig. 6.8 Feldspar pIRIR-225 single-grain dose distribution for sample NCL-7320043

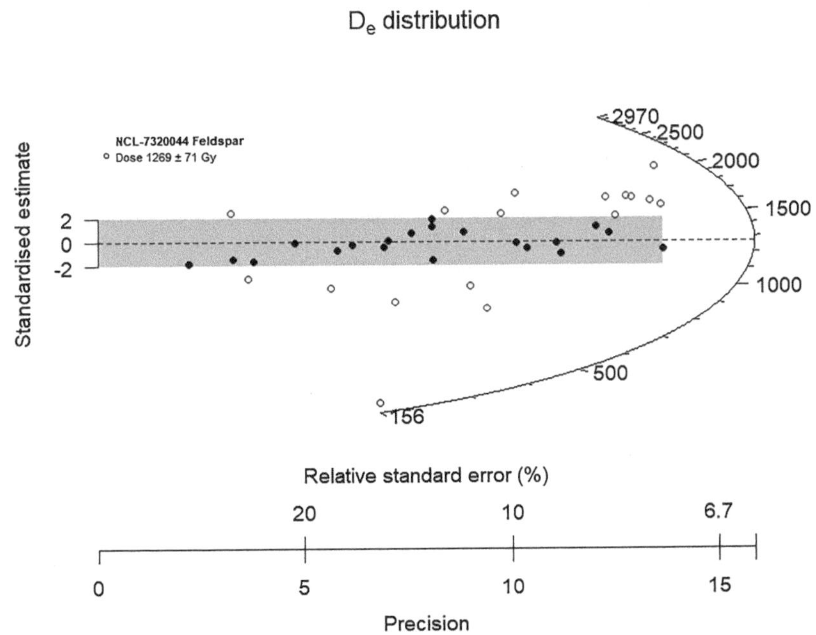

Fig. 6.9 Feldspar pIRIR-225 single-grain dose distribution for sample NCL-7320044

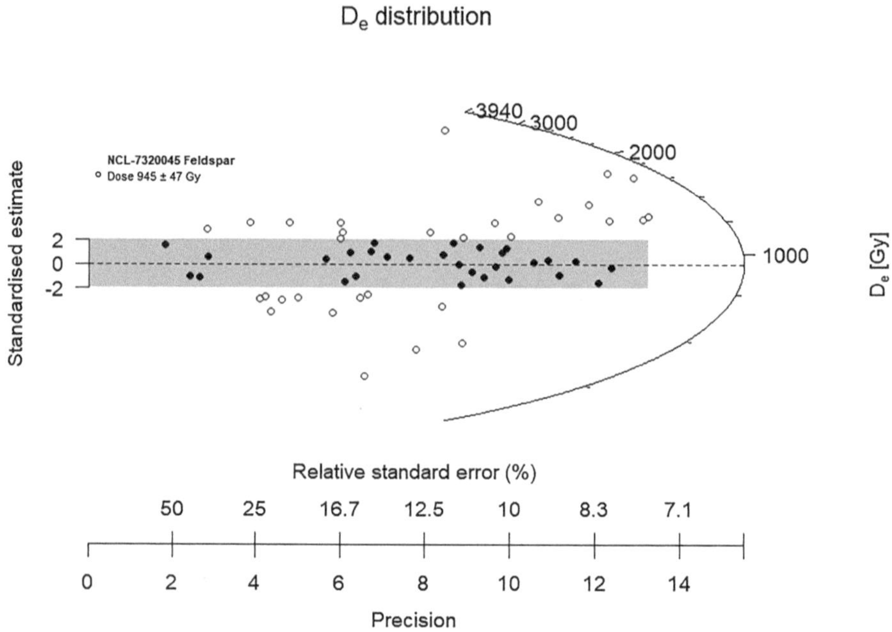

Fig. 6.10 Feldspar pIRIR-225 single-grain dose distribution for sample NCL-7320045

References

Brill D, Reimann T, Wallinga J, May SM, Engel M, Riedesel S, Brückner H (2018) Testing the accuracy of feldspar single grains to date late Holocene cyclone and tsunami deposits. Quat Geochronol 48:91–103. https://doi.org/10.1016/j.quageo.2018.09.001

Galbraith RF (1990) The radial plot: graphical assessment of spread in ages. Int J Radiat Appl Instrum (Part D) Nucl Tracks Radiat Meas 17:207–214

Reimann T, Román-Sánchez A, Vanwalleghem T, Wallinga J (2017) Getting a grip on soil reworking—Single-grain feldspar luminescence as a novel tool to quantify soil reworking rates. Quat Geochronol 42:1–14. https://doi.org/10.1016/j.quageo.2017.07.002

Index

H. Ahokas et al., *Paleoshorelines and Their Relict Littoral Plants*, SpringerBriefs in Earth Sciences, https://doi.org/10.1007/978-3-032-17723-0

FSC
www.fsc.org
MIX
Papier | Fördert
gute Waldnutzung
FSC® C083411

Zeitfracht Medien GmbH
Ferdinand-Jühlke-Straße 7
99095 Erfurt, Deutschland
produktsicherheit@kolibri360.de